44th Vibration Institute Annual Training Conference 2021 and Society for Machinery Failure Prevention Technology Annual Conference 2021

Arlington, Texas, USA
14-16 July 2021

ISBN: 978-1-7138-6056-3

Printed from e-media with permission by:

Curran Associates, Inc.
57 Morehouse Lane
Red Hook, NY 12571

Some format issues inherent in the e-media version may also appear in this print version.

Printed with permission by Curran Associates, Inc. (2022)

For permission requests, please contact Vibration Institute at the address below.

Vibration Institute
2625 Butterfield Road
Suite 128N
Oak Brook, IL 60523-3415
USA

Phone: (630) 654-2254
Fax: (630) 654-2271

information@vi-institute.org

Additional copies of this publication are available from:

Curran Associates, Inc.
57 Morehouse Lane
Red Hook, NY 12571 USA
Phone: 845-758-0400
Fax: 845-758-2633
Email: curran@proceedings.com
Web: www.proceedings.com

VIATC
VIBRATION INSTITUTE ANNUAL TRAINING CONFERENCE & EXPO 2021

MFPT
ANNUAL CONFERENCE

JULY 14-16 | ARLINGTON, TX

CONFERENCE PROCEEDINGS

44th Annual Training Conference and Expo

***Live!* By Loews – Arlington, Texas**

www.vi-institute.org | 630-654-2254

TABLE OF CONTENTS

VI PAPERS:

MFPT PAPERS:

SELECTION OF CBM SYSTEMS – QUESTIONS TO ASK

Author: Ed Spence

SELECTION OF WIRELESS CBM SYSTEMS – QUESTIONS TO ASK

Ed Spence
The Machine Instrumentation Group
535 Washington St, Dedham MA 02026
ed.spence@machineinstrumentation.com

Abstract: The Predictive Maintenance landscape has experienced an explosion of complexity in the last 10 years – IIoT, AI and machine learning, MEMS sensor technology, more sophisticated vibration signal processing and data engineering, edge processing – presenting the reliability engineer with both challenges and opportunities.

The new capabilities offer the promise of improving your asset uptime while saving you more of your most precious resource – your time. You would like to get your hands around all this new technology, but it's not easy to sort through all the hype, promises and complexity.

This talk will help you better understand some of the features and functions of Wireless CBM, technology dependencies and some of the things to think about when considering a system to test, pilot or deploy.

Keywords: Accelerometers; CBM; IIoT; MEMS; network; PdM; wireless; vibration;

Introduction:

This discussion will be restricted to wireless 'IIoT' systems promoted for industrial condition monitoring (CBM). The most commonly used measurand is vibration measured with an accelerometer, and is typically designed to monitor the vibration of rotating industrial machinery, measuring vibration, performing statistical calculations and other signal processing then transmitting signal waveforms and condition indicators for analysis, display and storage. That said, provision for other sensors commonly used for PdM and process control may also be supported by these systems.

Vendor Positioning:

Wireless CBM Systems defy easy characterization due to degrees of complexity that includes accelerometer performance, inclusion of other supported measurands, transmission distance and battery life, configuration and features, signal processing and data management, and network topologies.

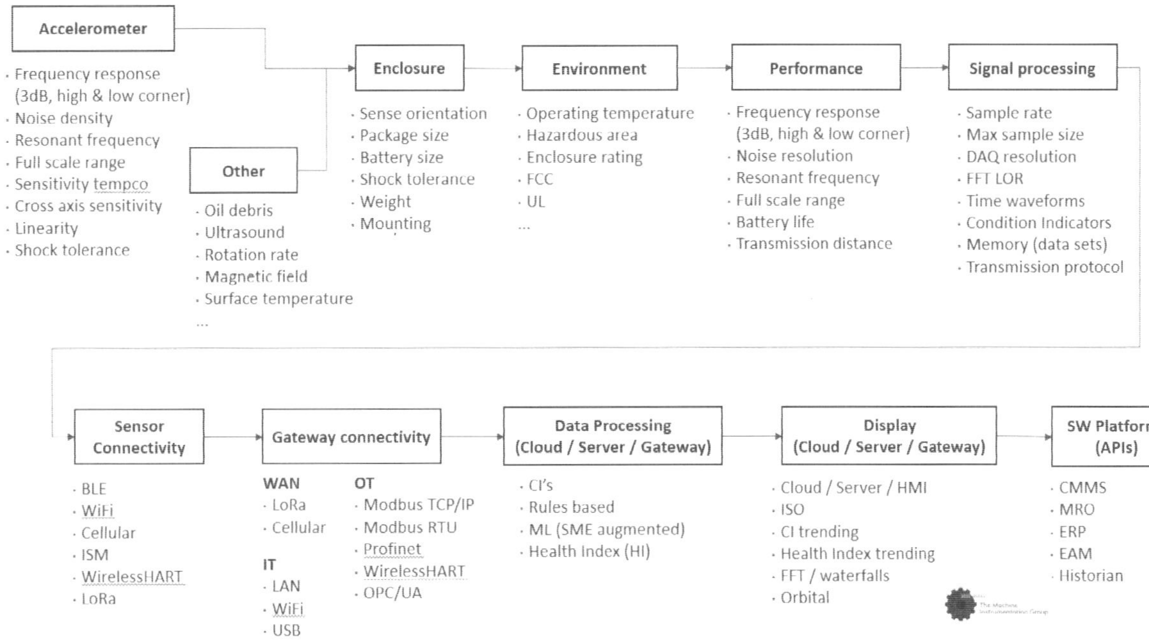

Figure 1. Wireless sensor systems for CBM represent not only the features common to IoT systems, but add degrees of complexity associated with vibration measurement and the various environments that the systems must operate in.

In addition, vendors come from very different backgrounds and degrees of PdM/CBM 'pedigree'. Traditional 'boots-on-the-ground' PdM service providers are acquiring, developing or recommending wireless systems to complement their existing services by adding wireless data collection - saving 'truck rolls', speeding up response times and increasing the frequency of measurements taken.

The maturing of enabling technologies such as Low Energy Bluetooth, low cost MEMS accelerometers and advanced data analysis techniques has attracted new entrants to CBM applications with 'IoT' competencies but less experience in PdM tools and techniques.

This has led to a rich mix of systems with various capabilities and price points. We'll look at some of the features of a system worth closer inspection before you try or buy.

Assessing the Options:

Perhaps the best place to start is by considering the flow of data – what kind of data would you want to see generated by the sensor, who will analyze it and where does it need to be sent?

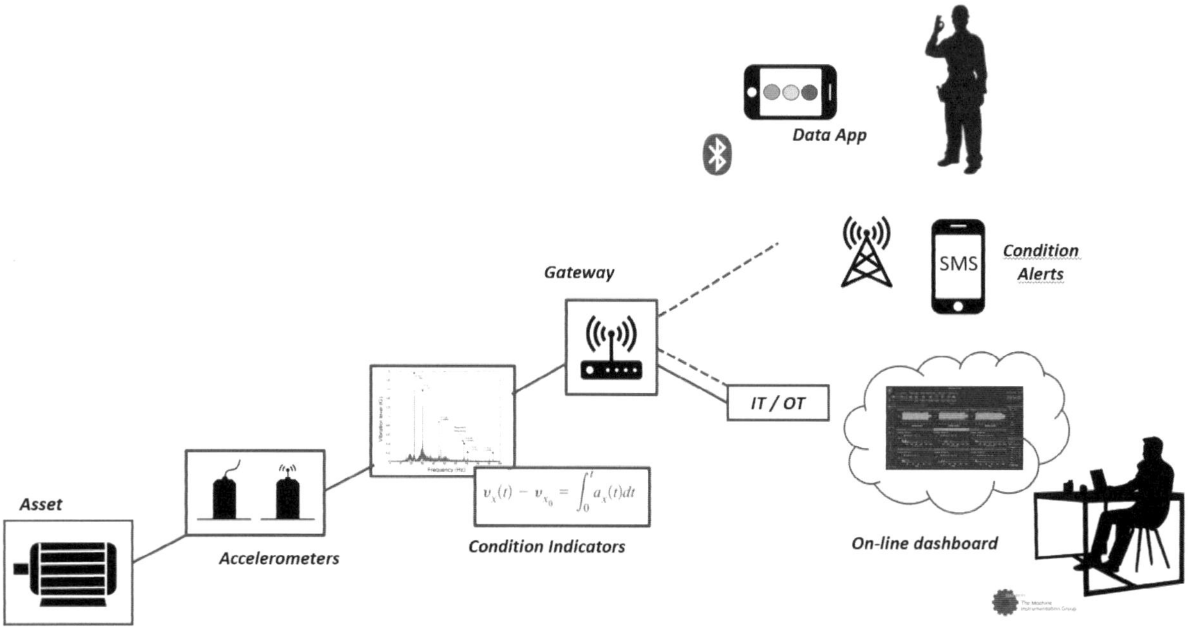

Figure 2. Selecting a wireless data collection system requires some thought regarding what kind of data you want, who will make use of it, where it needs to go and hence, what signal processing will be required.

You may simply want to monitor Condition Indicators such as velocity, flagging a trending change in the behavior of the asset and triggering an alert to go have a closer look. If this result is your aspiration, then you have plenty of options to consider and a first sort of the systems can be performed by reviewing the accelerometer specifications. Almost all vendors provide trended statistics such as RMS or Peak acceleration or velocity calculated in the gateway if not in the sensor itself.

If on the other hand you want higher frequency vibration data that supports more sophisticated bearing fault detection algorithms, then the options narrow but a preliminary sort can still be made by reviewing accelerometer frequency bandwidth.

Accelerometer Performance and Other Sensors:

It is no secret that MEMS accelerometers have helped to enable proliferation of wireless data collection. What is less well understood is what can be expected from the performance of the sensor, as vendors have used a wide variety of commercially available MEMS accelerometers. Experienced vibration analysts take for granted that the PZT sensor technology they are accustomed to using has the bandwidth necessary to support bearing diagnostic techniques, but with MEMS accelerometers this is not necessarily true. Closer scrutiny of the sensor technical specifications is required to determine whether the chip-scale accelerometer 'under the hood' of a particular wireless sensor has not only the bandwidth required but also the noise and full-scale range necessary to implement your chosen bearing fault detection algorithms.

Given the power, cost and other compelling features of MEMS accelerometers, many IIoT vendors provide triaxial acceleration sensing with modest sensor frequency response (ie 1kHz), often just enough to support calculation of ISO standard velocity levels. These systems may be more than adequate to 'raise a hand' to let you know that it's time to visit an asset for closer inspection, but saving you a lot of time by reducing a truck roll until it's really necessary. Many lower frequency mechanical anomalies can also be diagnosed in frequency domain with the spectrum plots provided.

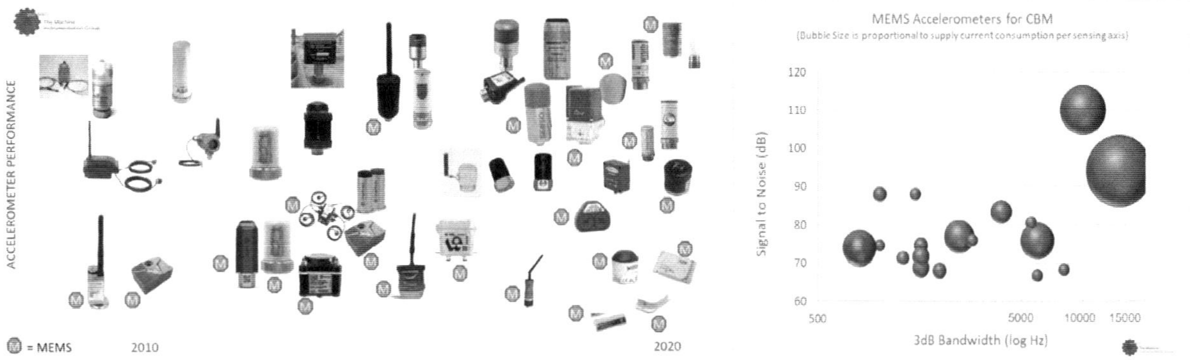

Figure 3. These positioning charts shows the wide range in performance for chip-scale (MEMS) accelerometers currently designed into wireless CBM systems. Lower bandwidth MEMS accelerometers will be triaxials and cheap, while the few higher frequency response sensors available will be more expensive and consume more power.

Beyond accelerometers, several vendors offer a broader suite of sensors in order to cover more measurands on the P-F curve. These could take the form of dedicated wireless sensors or alternatively a wireless 'hub' that provides analog input sockets for connection to traditional sensors. Supported sensors could measure pressure, oil debris, rotation rate or other measurands. The advantage of such a hub is that it can sport a bigger battery, offers placement flexibility for clearer transmission path and may even accept 24VDC or 120VAC power.

Network Connectivity:

A wireless CBM system has more than one connection option to consider - the connection between sensor and gateway, and that from gateway to the IT or OT network of the facility. As far as most of us are concerned, how the sensor connects to the gateway is of little interest, as long as the connection is robust and supports the signal processing strategy. Getting the data from gateway (or sensor) to the IT or OT network however, is more relevant for integration of the system into the target environment.

Vendors take different approaches to connection options of the wireless system, depending on their target applications or environment, the type and amount of data to be transmitted or their own competencies.

Vendors with a PdM or legacy CBM background often ensure that their system can communicate with the OT infrastructure through legacy protocols. For example, a vendor strategy may be to simply transmit Condition Indicators such as velocity RMS via WirelessHART (can be thought of as a wireless version of a 4-20ma loop) directly to the PLC or DCS. This option saves the cable cost of a traditional velocity 'transmitter'. In any case, it's important to note that vibration is unlike process control variables such as temperature and pressure, and the waveform must be reduced to a statistical indicator (scalar) that can be correctly interpreted by a PLC.

Most vendors, particular new entrants from an IoT background, offer gateway connectivity options such as Ethernet or WiFi for connection to the enterprise LAN, where a 'cloud'-based dashboard might reside on an 'on premises' server or a commercial platform external to the facility. Is remote access to the data from a geographically dispersed team or an external monitoring service necessary? If yes, then LAN access is a necessary option, and your IT department will want to be involved with this implementation.

Many wireless protocols originally developed with a particular set of applications or environments in mind have been adopted for CBM. Bluetooth, for example, originally for very short distance audio transmission, has migrated to higher data rates over longer distances, providing a low cost, low power protocol for connecting CBM sensors to your smart phone or a nearby gateway. LoRaWAN, well suited for 'smart city' applications transmitting small data packets over long distances, has also been applied to CBM systems in a few instances.

Table 1. Here is a partial summary of wireless CBM systems and supported gateway connections to the LAN or OT network.

PROTOCOL	DESCRIPTION	ABB (WiMON)	Acoem	Allied Reliability	Amazon	AMC VIBRO	Augury	Emerson	Erbessd	i-Care	IVC Technologies	KCF Technologies	Petasense	SEMEQ	SKF	Symphony Industrial.AI	SPM (Airius)	Ijssel (UpTimeWorks)	Waites Wireless	Machine Saver
MODBUS RTU	Serial bus ("Remote Terminal Unit") / RS485/RS232 PHY					Y								Y						
Modbus TCP/IP	MS Windows / IP address / Ethernet PHY		Y			Y		Y		Y										
WirelessHART	Transducer Protocol based on 802.15.4 (Emerson)	Y						IEC 62591*							Y					
EtherNet/IP	Common Industrial Protocol (CIP)																			
Profinet	Siemens								y (via Modbus TCP/IP)									Y		
OPC / UA	Real time Server / Client comms to PLC, DCS or PC	Y				Y		Y						Y			Y			
'Ethernet'	LAN			Y	Y	Y	Y	Y	Y	Y	Y	Y	Y			Y		Y		
WiFi	LAN						Y	Y	Y	?	Y	Y	Y			Y	Y	Y		Y
Cellular	SMS / Data Svc.; cellular modem			Y			3G/4G	Y	4G	3G/4G	GPRS	Y	Y			Y		LTE	Y	
LoRa (PHY) / LoRaWAN (protocol)	Low power WAN / ISM bands									Y								Y		

The Machine Instrumentation Group

Lastly, some systems by-pass the IT complications by provisioning with high data rate cellular services (LTE, 3G/4G/5G). Data oriented cellular protocols designed for IoT applications are also emerging, with narrower frequency bands and modest data rates. LTE-CAT-M1 and NB-IoT are examples of this class of service. Although cellular transmission may simplify deployment where the IT department is concerned, service has to be available in the facility location and will come with a subscription cost proportional to the required data rate.

Of course, any service which transmits to and stores data on an external cloud server, may raise concerns from management regarding data security. Check what provision the vendor has put in place for data security, such as level of encryption[1].

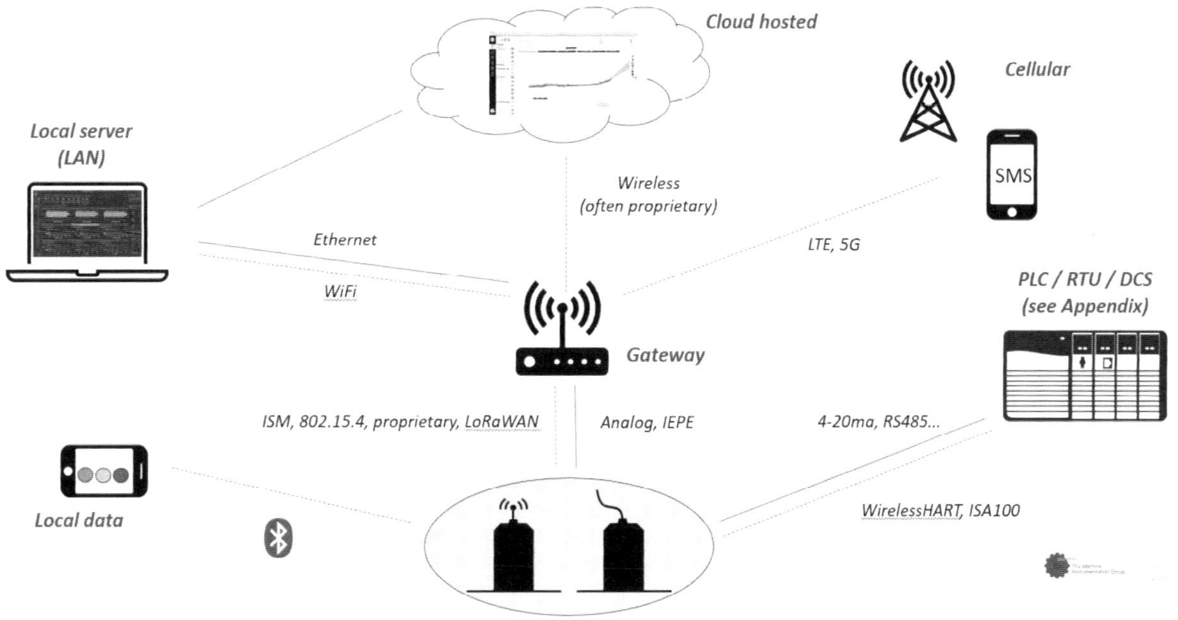

Figure 4. This infographic shows some of the common connectivity options provided by IIoT vendors serving CBM. Note that connections directly from the sensor will have different requirements than a LAN connection from the gateway.

Note that many vendors will transmit waveforms to their cloud dashboard for analysis by anyone on the maintenance team with login credentials. In this case a variety of tools and plots may also be provided for use by the analyst. More often than not, the data dashboard is located on a commercially available platforms such as Amazon Web Services (AWS) or Microsoft Azure. Connection to other software platforms such as the facility CMMS can be effected with an API software connection[2]. These are relatively straightforward software modules and many vendors will already have API's to talk to common OT or IT software platforms such as OSIsoft PI, and are busy writing new API's for other commercially available software packages and platforms commonly found in the industrial environment such as CMMS.

---

[1] There are two types - Encryption-at-rest vs. Encryption-in-transit. Many devices (e.g. medical) are demanding both. For example, if you only encrypt in transit (i.e. use the network's TLS encryption) and I steal the valve off your pipeline, is the data on it unencrypted? Encryption is built into the application, or the network protocol (which is more standardized). Either introduces some processing overhead consuming both time and power.

[2] Application Programming Interface, or software intermediary that allows two applications to talk to each other.

Monitoring and Analysis: Who will take primary responsibility?

If internal maintenance resources will maintain responsibility for monitoring and diagnosis then you must ensure that one of two features are in place. Either 1) you can easily get access to waveforms generated by the accelerometer for analysis on your own platform or with your own tools or 2) the cloud based dashboard provided by the vendor must have all the familiar tools, visuals and analysis features. Some vendors are reluctant to open their systems to the outside, preferring to sell a monitoring service, so be sure to ask about access to waveforms.

Many vendors provide remote monitoring and diagnostic services as well. Outsourcing the monitoring and diagnostics to the vendor may reduce costs of manual collection or simply alleviate a man-power shortage, saving time of your most valuable and skilled resources.

IIoT vendors take different approaches to data analysis, depending on their experience with asset health monitoring, their own signal processing or data analytics competencies and their Software development capabilities.

Traditional PdM service providers may have accrued valuable asset model data over the years, such that they can implement rules-based monitoring or diagnosis based on detailed knowledge of the asset physics. This approach is often promoted as 'AI', and may indeed use some supervised machine learning techniques, to adapt the rules to particular assets.

Several IIoT-oriented vendors to enter the market in the last ten years come with data science backgrounds, hence offer an 'AI' based approach to diagnostics. This approach favors multiple measurands as inputs in order to look for correlations and patterns between sensors on the asset.

With the emergence of 'data driven' approaches to analytics[3], data scientists can also extract features from any dataset, and work with an asset Subject Matter Expert (SME) to assign meaning to values and to help identify outliers, or data points that don't belong to a particular group or cluster of data. In this case, multiple contributing inputs can be modeled as a single health index or score, simplifying asset monitoring displays. Machine learning techniques are implemented to establish a baseline for the data trends, and alarms are set statistically in consultation with the asset SME.

[3] As opposed to classic vibration signal processing modelling the physics of the asset

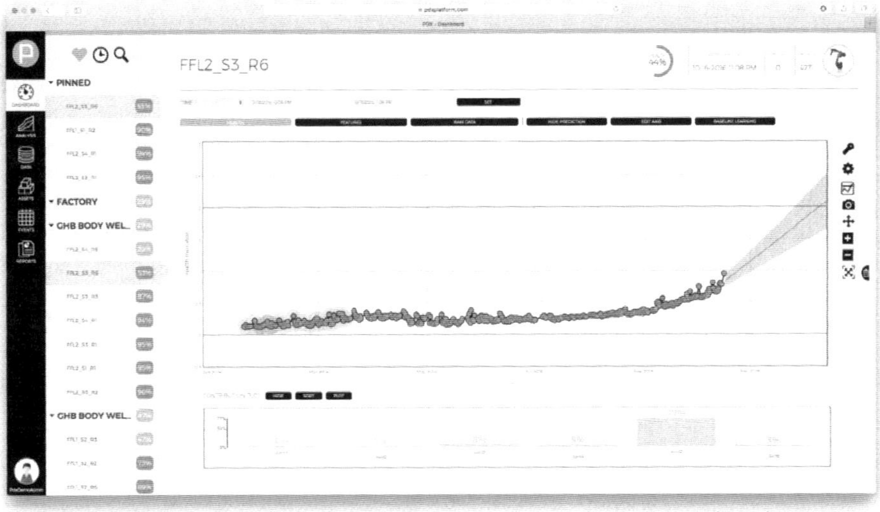

Image credit: Predictronics Corp.

Figure 5. An example dashboard trending a health index for automotive robotics. The index was developed using servo control data, and the component contributions are shown in the 'bar chart' below the trending graph. Note that for this asset, a robotic arm used in a manufacturing setting, Servo #5 is shown as the primary contributor to the index moving into the alarm state.

IIoT vendors without these competencies are generally offering standard Condition Indicators such as trended Peak or RMS acceleration or velocity, with an ISO based overlay of alarm levels. Conversion to frequency domain to provide a spectrum is also relatively common on vendor dashboards, sometimes with the ability to compare multiple spectrum plots in a waterfall chart.

Wireless CBM System Topologies:

Wireless CBM systems come in 4 basic topologies, each better suited for particular assets, applications and use cases.

A wireless 'junction box' extends the use of legacy PZT sensors with IEPE interface, making data from permanently installed sensors more accessible to maintenance vibration specialists. The junction box accepts cables supporting the standard IEPE interface, transmitting the data from multiple sensors to an IT destination. National Instruments and IMI promote products in this category.

The wireless 'hub' configuration has a similar function to that previously described, but in a smaller form factor and accepting inputs from a broader array of analog cabled sensors. These wireless hubs may be powered and provide more flexible system monitoring, accepting analog sensor inputs from process control sensors or tachometers as well as cabled accelerometers provided by the vendor. Dynapar, Waites Wireless and a multi-sensor hub configuration recently introduced by KCF Technologies fit into this category.

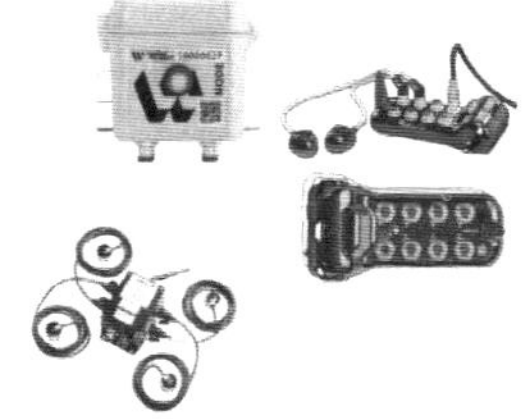

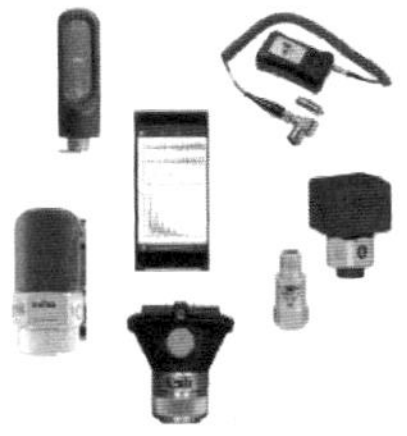

Several products on the market today offer to 'digitize' the route-based data collection process with a hand-held sensor which transmit vibration data to your phone or tablet. These sensors are not intended for permanent installation and are designed for continuous manual use by the specialist, hence have limited battery life. Vendors providing such devices include SKF, Ijssel, GTI Predictive. Erbessd and CTC.

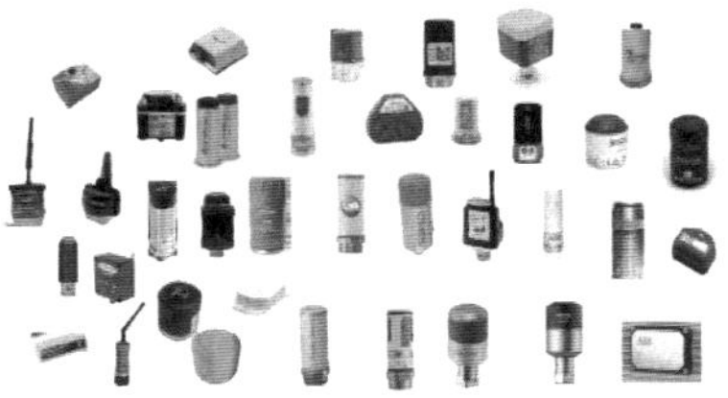

Finally, the last category of wireless sensors are those which many envision when they hear or read "IIoT". This class is the integrated wireless accelerometer transmitting waveforms or condition indicators to a corresponding gateway, which in turn forwards data over a variety of transport options and protocols. With shapes as small as a pill bottle, and as large as a soda can, these sensors are mounted directly on the asset for permanent, wireless data collection. The vendors supporting this category of product are too numerous to mention, while new vendors seem to appear almost monthly!

In Any Case, Battery Life:

Because transmission of vibration waveforms requires longer transmit time due to the higher amounts of data, battery life will be negatively impacted if this amount of data is transmitted too frequently (not to mention storage costs for the large quantities of data generated). Check the conditions cited for battery life of the system – less frequent measurements and/or transmissions can greatly prolong operation between battery changes. For this reason, 'smarter' systems might monitor condition indicators for an alarm condition before transmitting the full waveform. This can make sense from a system feature point of view because, although the data acquisition cycle is a significant part of the power budget, in general the radio transmission is the highest consumer of system power, such that reducing the portion of overall operation dedicated to data transmission will have the biggest impact on battery life.

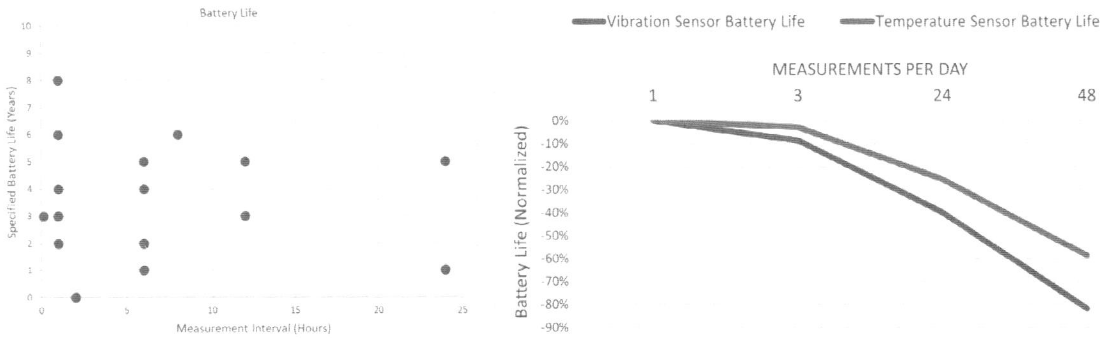

Figure 6. A plot of specified battery life, normalized by measurement interval for select wireless vendors, along with modeled radio performance from a wireless sensor design. Note that in general, a vibration waveform will have a bigger package size, hence transmit time, then a scalar such as temperature.

CAPEX vs OPEX

Finally, how do you want to pay for all this new technology? These systems have a wide range of prices, paid per asset or sensor node, ranging from single digit $'s to $10's/asset/month. Some offer subscription only, with one going as far as taking complete ownership for the health of your assets with limited liability insured by an underwriter. In an age where we pay a subscription for most services we consume (mobile, internet, media content), an 'aaS' (as a Service) option may be easier for your CFO to sign off as an OPEX expense.

To further complicate things, many PdM service providers offer wireless data collection as a service as part of their traditional 'boots-on-the-ground' data collection, diagnostics and reliability consulting services.

Summary

Unfortunately, the one question most difficult to answer is - how easy to install and configure any of these systems in your plant? Many vendors tout their systems as 'plug-and-play', but talking to service providers we have found that assessing the options *in-situ* is not always straightforward. Every facility has its own unique set of challenges for wireless transmission and data processing, such that a pilot or field test is the only real way to uncover these issues. Third party services are available to help with this...

All things considered it is an exciting time to be in maintenance, with many new tools and capabilities available for addition to the maintenance toolkit, with all the potential to save time and money.

I'm often asked the question 'which system is the best of the bunch'?

This is a little like asking 'I'm visiting Boston for the first time, and would like to impress my client, would you recommend the best restaurant in the city'? My answer has to begin along the lines of 'well, it depends – let me ask you a few questions about your tastes and budget, and we'll narrow the options from there...'.

Ed Spence is the Founder and Managing Director of The Machine Instrumentation Group, a consultant representing a network of contract engineering service providers with expertise in Condition Monitoring, sensor design, signal processing and data engineering.

Previously, Ed was the Marketing Manager for Analog Devices MEMS Sensor Technology Group, where he defined the MEMS accelerometer roadmap for vibration based Condition Monitoring, winning an Innovation Award at Sensors Expo 2017 for the flagship

ADXL1002 MEMS accelerometer. Ed has been engaged in the development of Condition Monitoring sensor solutions since 2008. Ed has published or presented on the subject of sensors for CBM on numerous occasions.

Ed serves on the Board of Directors for The Society for Machinery Failure Prevention Technology, and is a member of the Vibration Institute, The Society of Maintenance and Reliability Professionals and The Society for Prognostics, Health and Maintenance. Ed also volunteers for non-profit IBEC Ventures, providing business consulting to small business start ups overseas. Ed holds a B.S.E.E from University of Massachusetts, and a certificate in Strategy and Innovation from MIT Sloan Business School.

VIBRATIONS BEFORE AND AFTER THE MODIFICATION OF WATER SEPARATOR SILENCER

Author: Paul Liang

Vibrations Before and After the Modification of Water Separator Silencer

Paul Liang

Durr Universal, Inc.
Clean Technology Systems
Engineering
1925 Highway 51 - 138
Stoughton, WI 53589

After a weld failure occurred, the water separator silencer was reinforced at the inlet box and re-padded around the chamber's mid-section. After repair, we took the vibration measurements again at the same points. The vibrations were significantly reduced: the average vibrations were reduced about 5- 10 times and meets the preset target of 5 times reduction. The maximum reduction was 15 times at the flat panel of the silencer entrance.

Roots blower; Pulsation; Silencer vibration; Sand process

Introduction:

The water separator in the sand process is used to separate the air-water mixture before the mixture exits the separator to the atmosphere. It uses an off-set inlet box to create a swirl flow. The centrifugal force will throw out the water droplets on the separator's wall. The air will leave the separator's top and the droplets will drop down on the separator's bottom to a sewer system, as shown in Fig.1.

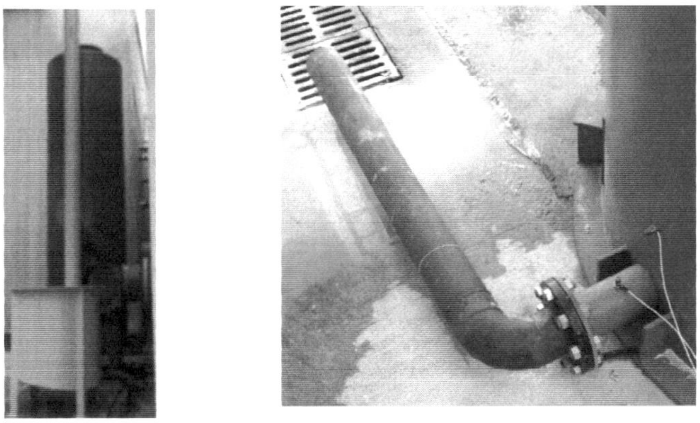

Fig.1 A typical water separator in sand process. Left) Air left from separator's top. Right) water left from separator's bottom pipe.

The water separator is acoustically like reactive silencers, which are used in a typical Roots blower. The separator's sudden expansion at the inlet and the sudden contraction at the outlet works as a single chamber muffler. The pulsation reduction from the silencer is predicable [1]. The separator is also like the reactive silencers in that the pulsations generated by the blower can be amplified by acoustical natural frequencies of the chamber if it is not designed properly. The pulsation can also excite the shell's structural natural frequencies, resulting in high vibrations and excessive noise levels. A finite element analysis (FEA, [2]) was used to calculate the silencer's structural modes to ensure that the structural modes were located out of the high pulsation peaks. The FEA results will not be included in this paper to limit the discussion.

The historical failures occur at the joint line of the silencer's inlet square box. This box has a large flat panel with either a cross bar or a circle ring to reinforce the panel as shown in Fig.2 below. However, these reinforcements were not strong enough to resist the strong gas pulsations in the long run. The blue silencer in Fig.2 will be discussed in this paper.

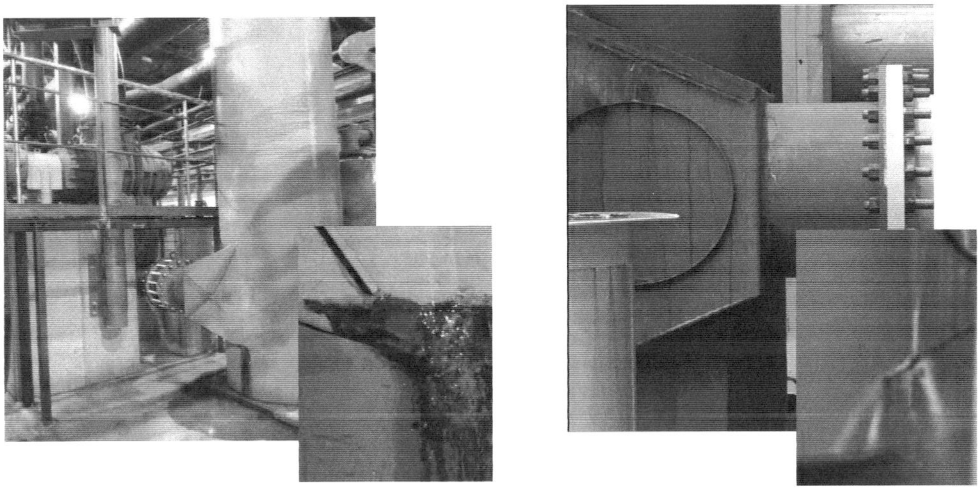

Fig.2 Failed water separators. Water gushed out through the inlet box's bottom edge of the broken weld seams.

Test Setup and Measurement:

We visited the failed water separator silencer on the morning of a cloudy/ light rain day. The temperature was ~70 °F. The blower connected to the silencer through a φ24" horizontal pipe, as shown in Fig.3. There was a flexible joint between the blower and pipe. The water gushed out through the bottom corner because of the leakage of the inlet box. There were other leakages, but they had been repaired before the visit.

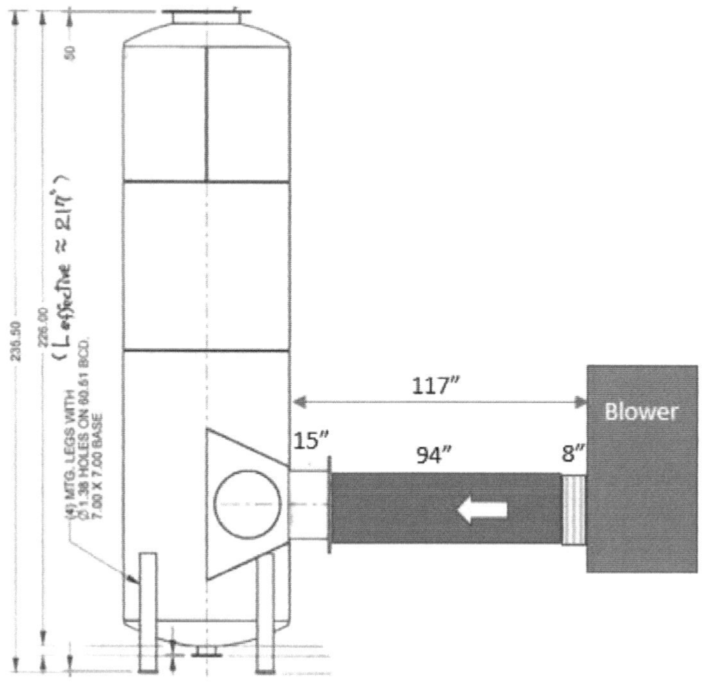

.**Fig.3 The silencer setup.**

A 2-lobe Roots blower, Howden Whispair 2050VJ, with a 5.9 psig discharge pressure and 16,500 iCFM, was used in the sand manufacturer. The blower ran at 723 rpm and was driven by a 700 HP electric motor. The gas temperature inside of the silencer was estimated as 190 ^{o}F based on a similar size blower. The blower's fundamental pocket passing frequencies (PPF) was four times of the blower's speed. The first four PPF(s) have been calculated as 48.2, 96, 145 and 193 Hz, respectively. The PPF can be calculated as 4x(723/60)= 48.2 Hz. Each shaft revolution releases four air pockets.

The internal gas pulsation level was not measured at this time, but its maximum level should not exceed 8% (or 1.6 psi) of the discharge's absolute pressure per manufacturer's guideline [3]. The inlet pulsation should not exceed 5% of the inlet's absolute pressure. The flange to flange distance was 94". The distance between blower and silencer was 117", as shown in Fig.3. The silencer dimensions are: φ24" inlet, φ60" chamber, L=217", with ¼" plate thicknesses on both the inlet box and chamber wall.

Two single-axis accelerometers, PCB 320C03 with 10 mv/g, were used to measure the inlet box and chamber shell vibrations. A tri-axial accelerometer, PCB 356B21 with 10 mv/g, was used to measure the legs and the inlet pipe in three directions as shown in Appendix A. The vibration levels at different locations are listed in Tab.A as well.

Analysis:

This was a "forced" vibration failure. The pulsations always created vibration peaks regardless of the chamber dimensions, as shown in Fig.4. The blower created airflow pulsations at 48.2 Hz and its harmonics. The highest peak, 2xPPF, was 8 g (root-mean-square g or rms g) in Fig.4. The p2 of the front plate of the inlet box had the strongest overall vibration acceleration, 14.5 g (overall rms g), as shown in Tab.A in Appendix A.

The 2xPPF, 96 Hz, should be from the inlet pipe's acoustic natural frequency between the blower and the silencer (see Equation 1 below). The 3xPPF should be from the silencer chamber's cross-wall and longitudinal modes (see red cells in Fig.6). These will be discussed later. The supported legs (p10 to 13) seemed to vibrate with much lower levels (< 0.5 g).

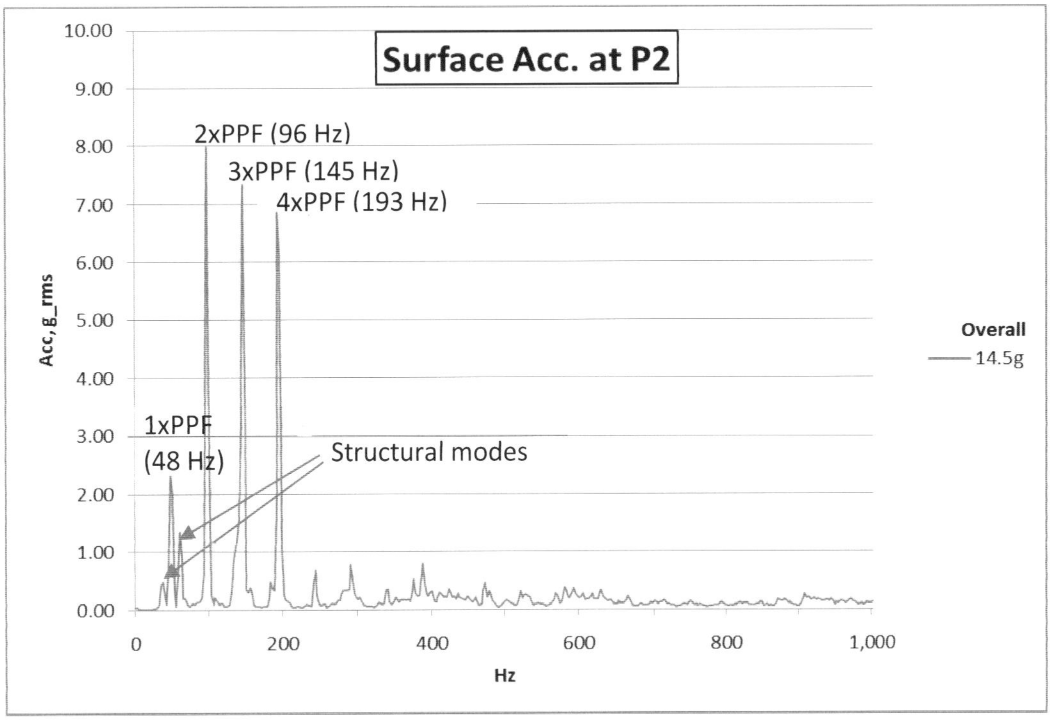

Fig.4 Measured vibration levels at inlet box (p2 in Appendix A). The 2x, 3x and 4xPPF have very high pulsations.

The peak accelerations were about 20 g as shown in Fig.5. The peak g is $\sqrt{2}$ higher than the rms g (=14.5*1.414). The instantaneous vibration peaks could be as high as ±40 g in the time-domain plot (Fig.5). The front plate needs to be reinforced so that the highest peak at 2xPPF will not exceed 1.6 g (or 5 times reduction). Typically, the peaks less than 1 g should not have structural failure; greater than 10 g will have short fatigue life.

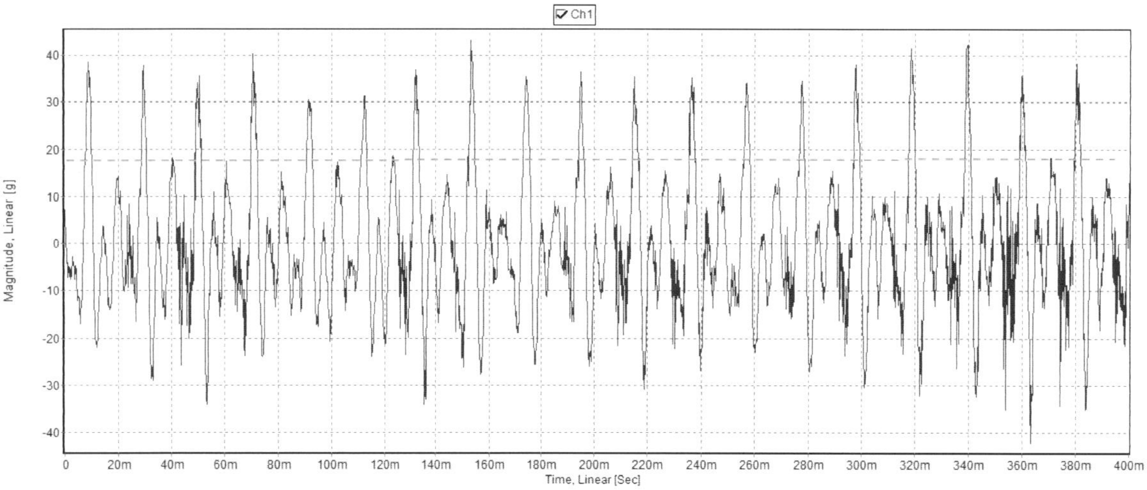

Fig.5 Time waveform at p2 (front plate of the inlet box). Most of the peaks were about 20 g. Some peaks could be as high as 40 g.

The section of the pipe between the blower and silencer can set up acoustic resonances. To avoid these resonances, certain pipe lengths must be avoided. For a pipe closed at one end (blower side) and open at the other (silencer side), the acoustic resonant lengths are given by [4],

$L = mc/(4f)$, $m = 1, 3, 5, \ldots$ (1)

L= pipe length in meter

f= PPF in Hz

c= velocity of sound in m/sec

The c is 381 m/s @190 °F. Therefore, the pipe length, 2.97 m (117”) created the acoustic mode 96 Hz at m=3, which aligned with the 2xPPF, 96 Hz very well.

The 1st cross-wall mode, 147 Hz and the 4th longitudinal mode, 138 Hz, were within the 5% range of the 3xPPF, 145 Hz, as shown in Fig.6. Although the 2nd cross-wall mode, 243 Hz aligns well at 5xPPF, 241 Hz, it is not critical. Typically, the first four pulsations are critical, and the chamber acoustic modes should avoid those peaks. The 1st cross-wall mode is more critical than the 2nd cross-wall mode. The reasons for the high 4xPPF, 193 Hz in Fig.4 are currently unknown.

Roots Blower Only

	Inputs:	
Blower RPM:	723	
# of Lobes:	2	
Gas:	air	
Operating Temp., F:	190	
	Silencer	Tube
ID, in:	60	24
	1st Chambe	2nd Chamber
Effective L, in:	217	10

(Please see Example 1 for the effective length derivation)

PPF (Excitation)	Hz
1st	48
2nd	96
3rd	145
4th	193
5th	241
6th	289

Resonance Criterion= 5%

Cross-Wall Modes (Receiver)

	1Silencer	Nearest PPF	2Tube	Nearest PPF
	Hz	Hz	Hz	Hz
1st	147	145	368	
2nd	243		607	

Longitudinal Modes (Receiver)

	1st Chamb	Nearest PPF	2nd Cham	Nearest PPF
	Hz	Hz	Hz	Hz
1st	35		750	
2nd	69		1500	
3rd	104		2250	
4th	138	145	3000	
5th	173		3750	
6th	207		4500	

Fig.6 Roots blower acoustic resonances prediction. The yellow area is input data and red cells are high risk modes.

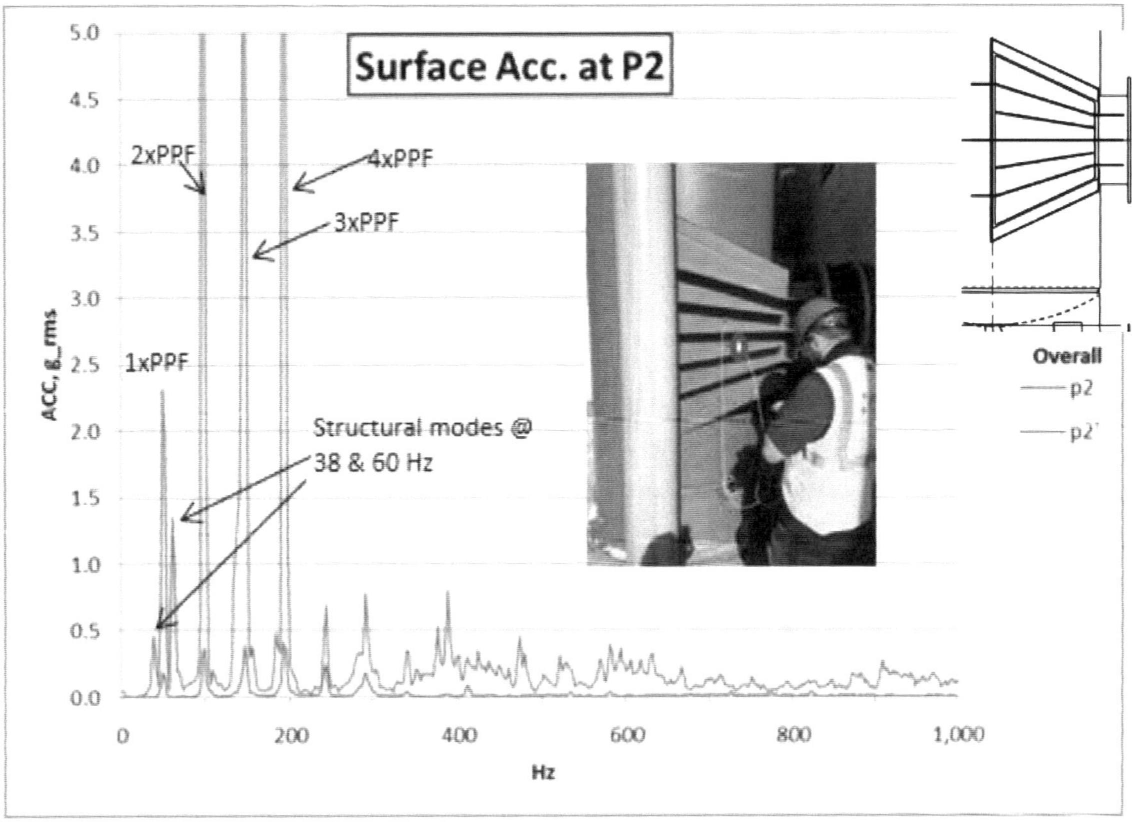

Fig.7 Vibrations at entrance panel, p2. Blue and red lines are before and after the modifications. The structural modes at 38 & 60 Hz were disappeared (red line).

To simplify the repair, we decided to only reinforce the inlet box. We used the parallel ribs (Fig.7) without changing the chamber and inlet box's plate thickness to drive the plate modes above 200 Hz to avoid the first four PPF pulsations. The rib greatly increased the plate's stiffness. Therefore, the plate vibrations were reduced accordingly. The replacement silencer was re-padded around the chamber's mid-section while the original silencer was not. The replacement silencer greatly reduced vibration levels, as shown in Figure 7. The overall vibration at p2 in Tab.A was reduced from 14.5 g to 0.93 g which was a 16 times reduction.

It is interesting to note that the structural modes of the replacement silencer at 38 & 60 Hz were complete gone because of the modal frequency shifted. The pulsation peaks were not shifted because they were "forced" vibration peaks (not the modes). Actually, the vibrations in entire frequency range were much lower because of the plate stiffness increase.

Discussion and Conclusions:

The water separator works as a single-chamber silencer. The chamber diameter and the length should be carefully designed so that it will not align with the critical blower's PPF(s). The inlet box needs extra reinforcement to resist strong gas pulsations. Typically, the first four peaks are high.

The vibrations at the chamber's top were lower than the adjacent area of the inlet box. These vibrations seemed to be from the inlet box's strong vibration. In the Tab.A in Appendix A, the silencer's top vibrations at p14-15 (3.3 g to 4.0 g) were lower than adjacent points of the inlet box at p5-9 (4.7 g to 6.2 g). It also can be seen by the replacement silencer's vibration levels. The vibration levels at all the points on the silencer shell were reduced significantly (Tab.A) because of the reinforced inlet box. The shell thicknesses in both silencers were the same.

References:

1. P. Liang, "Blower's Pulsation dampener using reactive silencers." Inter-noise 2018, Chicago, IL. August 26-29, 2018

2. J. Dorgan, "FEA Analysis on a failed water separator." Internal engineering analysis, Durr Universal. April 2016

3. Howden, "Roots RAS and RGS Whispair Blower – 10"-20" Lubricated, Installation Operation & Maintenance Manual." (From Google search)

4. A. Brummer, "Causes and effects of pulsations in compressor systems." Technische Universitat

Appendix A: Accelerometer Locations & Vibration Levels

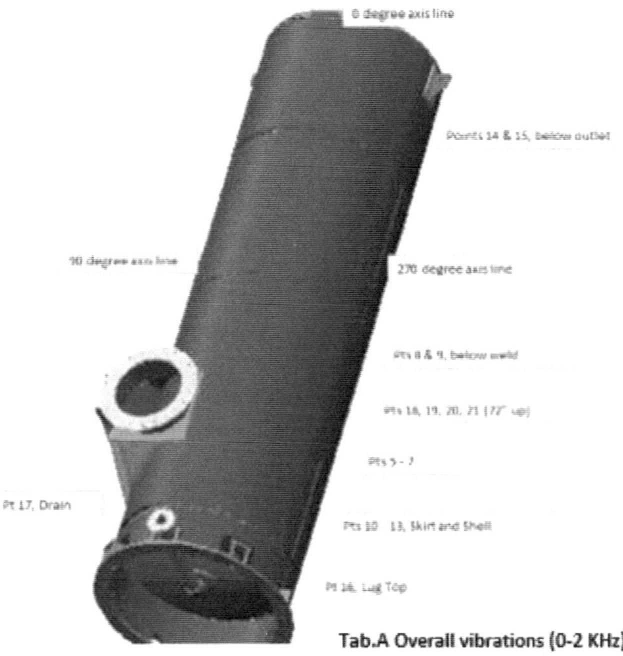

Location	Sensor location	Sensor 1	Sensor 2
0	Inside building after EJ	right	left
1	Inlet collar at near edge	down	up
2	Flat ribbed section from center	down	up
3	Top of inlet (at edge)	on top edge	at corner
4	Bottom of inlet (at edge)	on bottom edge	at corner
5	90 degree center, 60" up at CL	top	bottom
6	180 degree center, 60" up at CL	top	bottom
7	270 degree center, 60" up at CL	top	bottom
8	90 degree below middle weld	bottom	top
9	180 degree below middle weld	bottom	top
10	45 degree skirt and shell	shell	skirt
11	135 degree skirt and shell	shell	skirt
12	225 degree skirt and shell	shell	skirt
13	315 degree skirt and shell	shell	skirt
14	90 degree 18" below outlet head	bottom	top
15	180 degree 18" below outlet head	bottom	top
16	on lug top plate	225 degree	315 degree
17	Pipe outlet	top	skirt
18	0 degree, 72" up on shell	right	left
19	270 degree, 72" up on shell	right	left
20	180 degree, 72" up on shell	right	left
21	90 degree, 72" up on shell @ corner	right	left

Tab.A Overall vibrations (0-2 KHz) before and after modifications.

Point	Before	After
0	0.39 g (y)	0.63 g
1	0.62 g (y)	0.93 g
2	**14.5 g**	**0.93 g**
3	3.5 g	0.60 g
4	6.1 g	0.85 g
5	5.8 g	1.04 g
6	6.1 g	0.79 g
7	4.7 g	0.66 g
8	6.2 g	0.81 g
9	5.4 g	0.79 g
10	0.25 g (y)	0.58 g
11	0.31 g (y)	0.87 g
12	0.33 g (y)	0.77 g
13	0.39 g (y)	0.67 g
14	3.3 g	1.33 g
15	4.0 g	1.70 g
16	-	0.14 g
17	-	0.22 g
18	-	1.29 g
19	-	0.71 g
20	-	0.85 g
21	-	0.99 g

GAS TURBINE VIBRATION CASE STUDY – IS IT IMBALANCE OR IS IT A RUB?

Author: Timothy S Irwin, P.E.

VIATC VIBRATION INSTITUTE ANNUAL TRAINING CONFERENCE & EXPO 2021
MFPT ANNUAL CONFERENCE

2021 CONFERENCE PROCEEDINGS

GAS TURBINE VIBRATION CASE STUDY – IS IT IMBALANCE OR IS IT A RUB?

Timothy S Irwin, P.E.
Mitsubishi Hitachi Power Systems Americas, Inc.
Gas Turbine Service Engineering - Principal Engineer
2287 Premier Row
Orlando FL, 32809
Timothy.Irwin@amer.mhps.com

Peyton Swan, P.E.
Brüel & Kjaer Vibro
SETPOINT™ Diagnostics Services Manager
12818 Wallingford Drive
Tampa, Florida 33624
Peyton.Swan@bkvibro.com

Abstract: The work that this case study describes was due to high vibration levels in a large frame gas turbine. Higher than desired vibration levels occurred after a major inspection was performed with replacement of all compressor and turbine blading. Initial vibration data analysis with balance shot response and subsequent inspection results and additional balancing efforts will be reviewed.

Keywords: Gas turbines; diagnostic; vibration analysis; rubs; imbalance; balancing

Introduction:

There are several thousand large frame gas turbines in operation in the United States. The technical concepts of gas turbines have been around for decades and there are many books and articles available that document the history and development of these machines. Technological advancements in recent decades have allowed the gas turbine to become the preferred land based fossil fueled power generation method for base load and load following duties today. The advances in metallurgy and technical concepts have allowed the large frame gas turbine to exceed 40% efficiency in simple cycle mode and to exceed 60% efficiency in combined cycle mode (including a heat recovery steam generator and steam turbine bottoming cycle). The quick response and load following capabilities of the most recent designs are a very good match for balancing load demands as renewable energy power sources become a larger percentage power source for the electrical grid.

Given the advancements made in these machines, they are still large turbo-machinery that can incur the typical problems associated with all mechanical rotating machinery. Imbalance and rubs are a large portion of the typical issues that affect operational behavior and reliability. This paper will discuss a case study in which both these conditions were affecting a machine during initial operation after a maintenance outage.

This case study involves work performed on a large frame gas turbine rated over 150MW. The machine was installed in 2008 and had accumulated approximately 400 starts and over 80,000 hours of service at the time of the major inspection. The case study will focus on the gas turbine vibration signature and will not discuss or review the vibration signature of the generator.

Gas Turbine Components and Outage Work Scope:

Figure 1 below shows a general cutaway of a large frame gas turbine. Figure 1 will be referenced for the discussion of the issues that occurred and resolution of those issues. The figure referenced shows the air inlet to the left and the exhaust to the right.

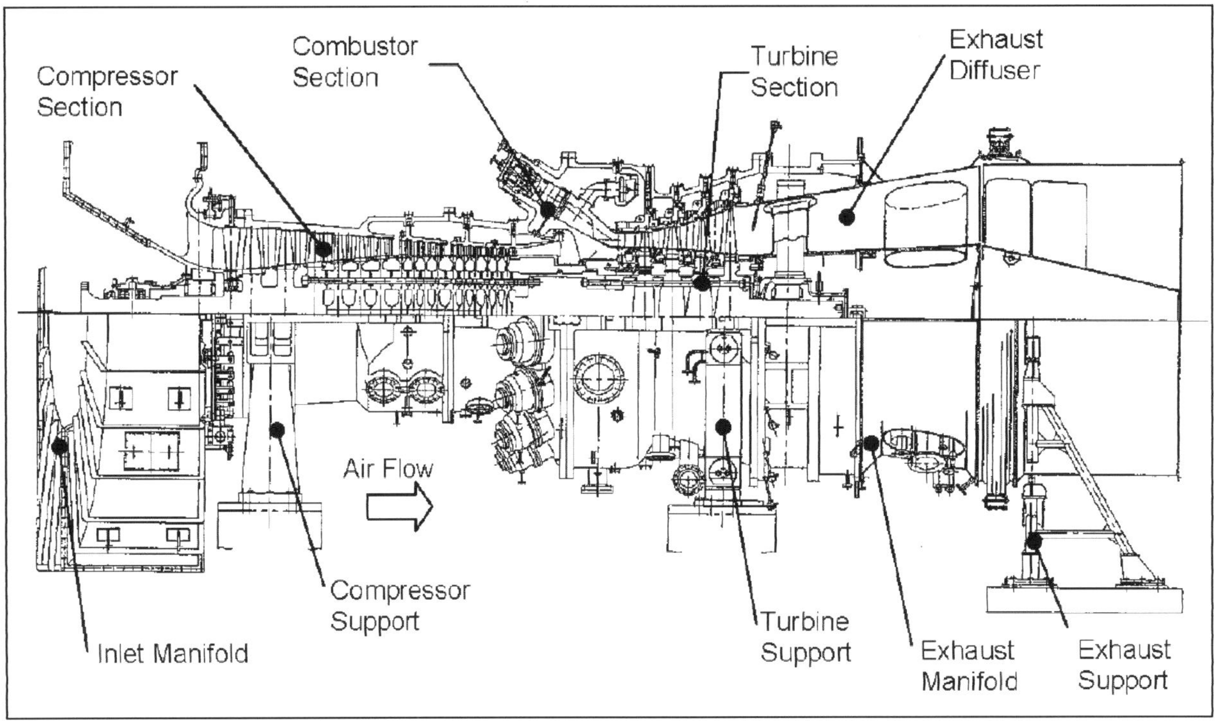

Figure 1 – Longitudinal

The outage work scope prior to the vibration data acquisition and analysis efforts was a major inspection and upgrade which included rotor uncoupling and removal. The overall scope included inspection of all components including bearings and replacement of combustor parts, turbine parts, and partial replacement of compressor parts.

Initial Startup Data Analysis:

Figure 2 below shows the frame of reference for the vibration analysis. Documenting the frame of reference for vibration analysis efforts is critical for future reference and when communicating with different groups to ensure everyone understands the communication intent of the analyst.

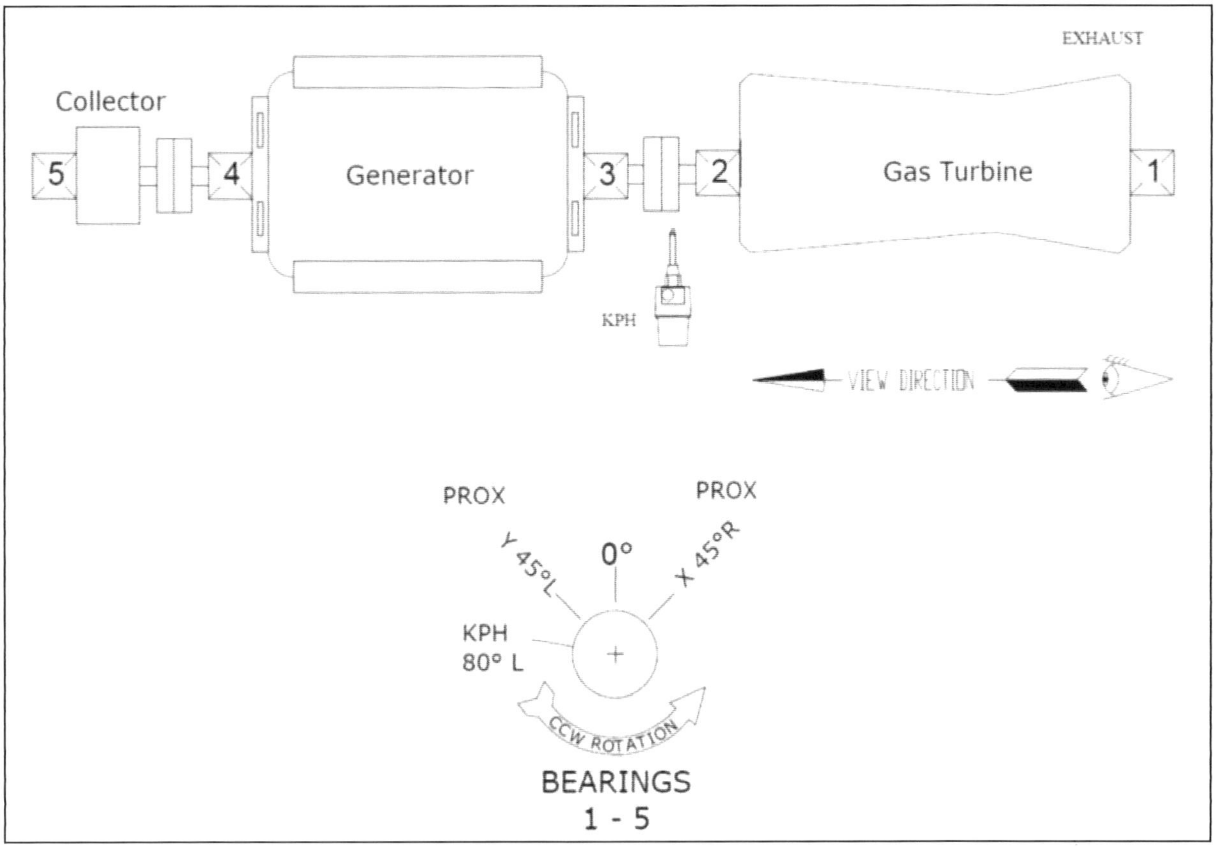

Figure 2 – Frame of Reference

The initial startup and operation of the unit exhibited vibration amplitudes reaching approximately 9.5 mils on the #2 bearing X channel. The #2 bearing Y channel amplitude reached ~ 7.5 mils. The #1 bearing X/Y channels reached approximately 5.3/3.3 mils respectively. The #2 bearing amplitudes were much higher than recommended and after a short operational period the machine was shutdown to allow for vibration analysis and development of a formal recommendation for investigative/corrective actions.

Figure 3 below shows the overall amplitude trends during the initial operational period. It was noted that while the amplitudes at the #2 bearing were much higher than the #1 bearing, there was also 'separation' in the X/Y channel amplitudes at each bearing with the X channel being higher at both bearings.

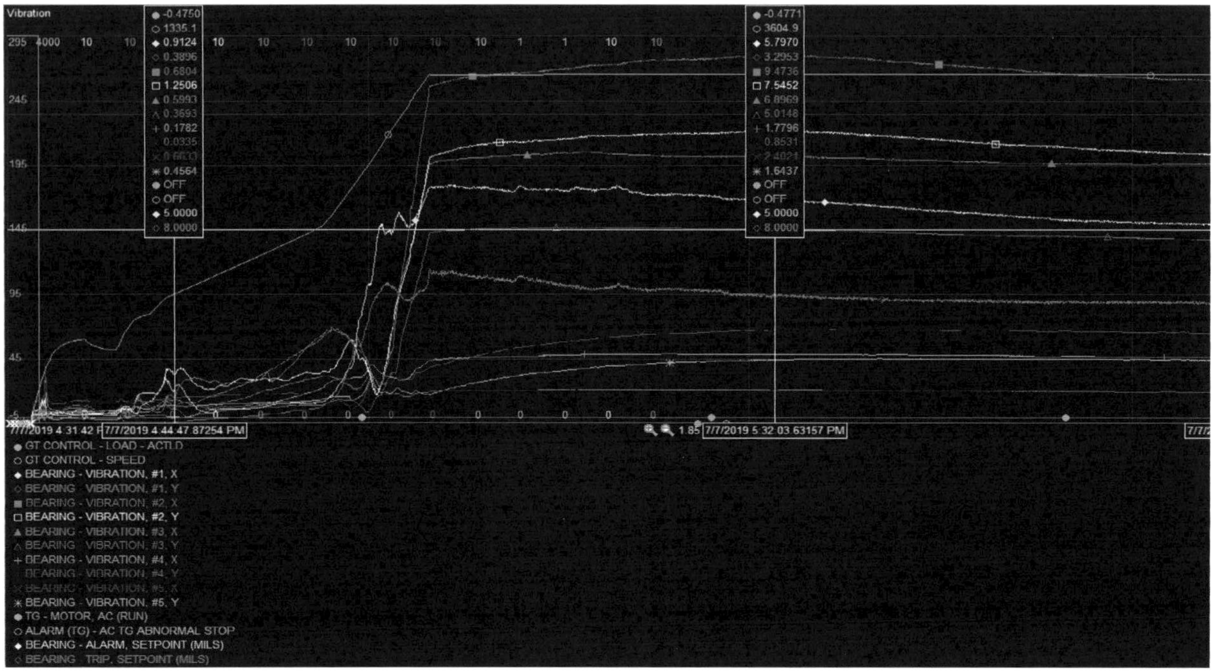

Figure 3 – Overall Amplitude Trend Values

Given the overall trend amplitudes during operation, review of the dynamic vibration data during the startup/operation/shutdown was performed. Evaluation of the startup Bode and Polar plots (Figures 4 and 5 below) indicated an unusual behavior. The #1 bearing never appeared to fully clear the 2nd shaft critical and the #2 bearing did not clear the 2nd shaft critical.

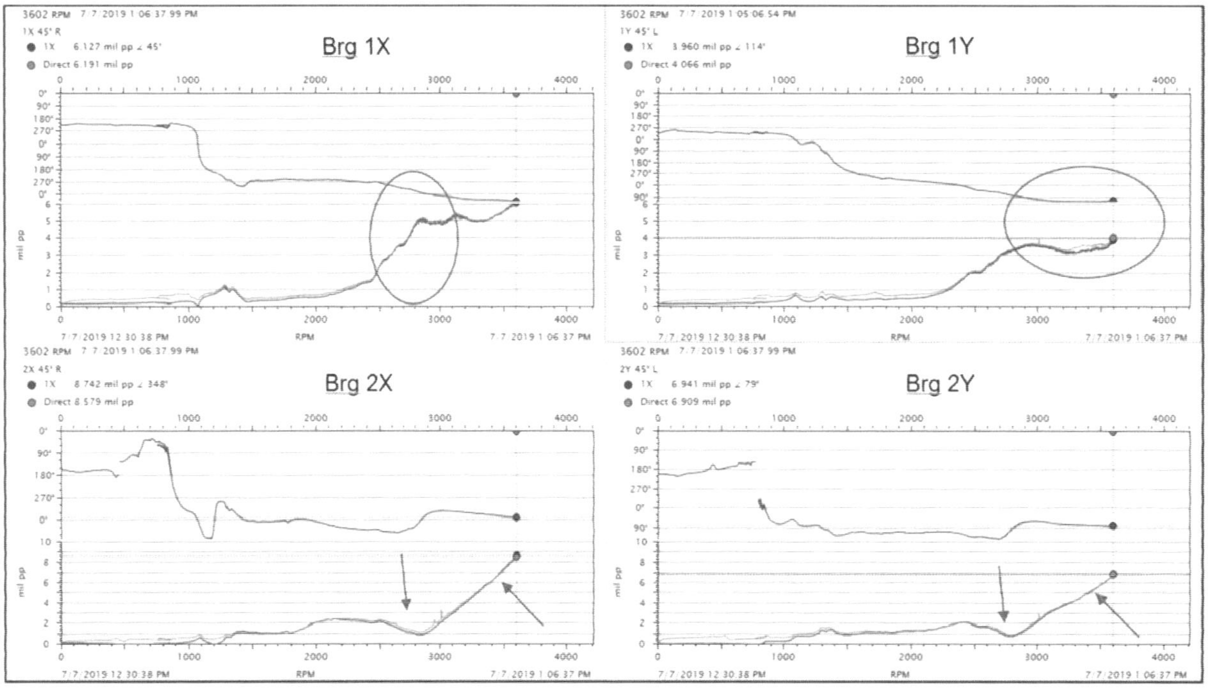

Figure 4 – #1 and #2 Bearing Bode Plots

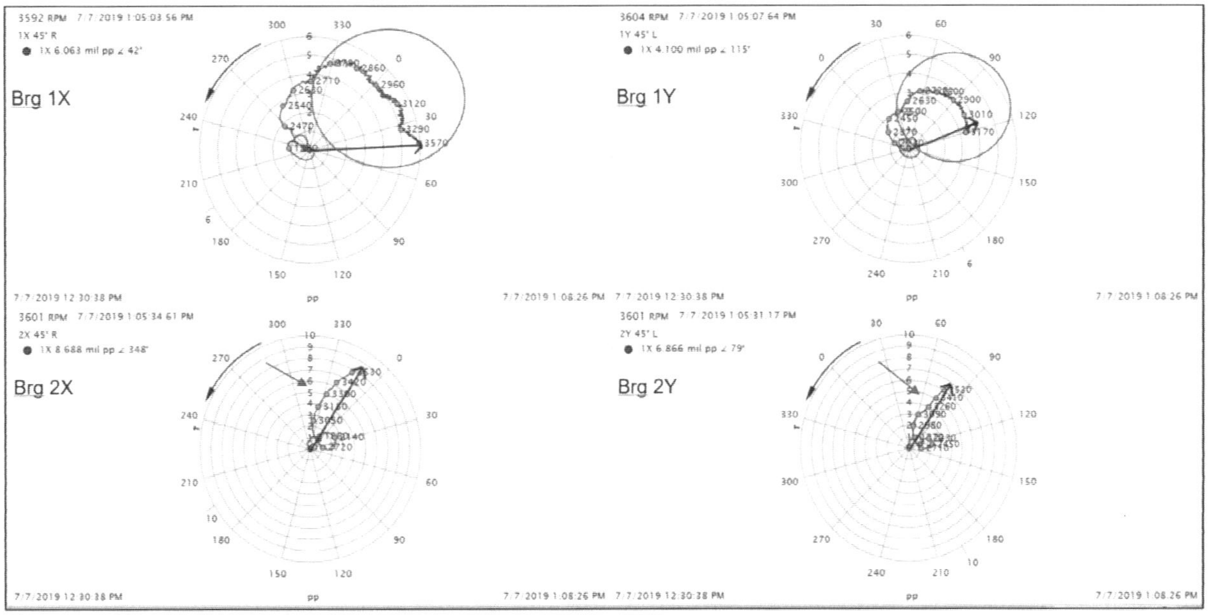

Figure 5 – #1 and #2 Bearing Polar Plots

The Bode and Polar Plots clearly indicated an abnormal behavior with a phase rotation of the 1X vector at the #1 bearing (but very little amplitude change) after the 2nd shaft critical and an amplitude climb at the #2 bearing with very little phase change after the 2nd shaft critical.

An exhaust plane balance shot was initially installed which resulted in very little response in the vibration signature. Then a center plane balance shot was attempted also resulting in very little vibration signature response. The results of the balancing attempts were that the signature still indicated an inability of the machine to clear the 2nd shaft critical.

Review of the installed balance shot influence was ~ ½ of the expected influence, this condition was also indicating that additional influences were affecting the operation of the machine.

At that time it was recommended that the machine be disassembled for inspection to determine the cause of the increased stiffness (i.e. rub) that was causing the abnormal behaviors at both bearings and the apparent increased speed of the 2nd shaft critical.

<u>Re-Inspection Results:</u>

During the re-inspection of the machine minor rubs between shaft and stationary seals were noted at various areas in the torque tube and initial turbine stage. Harder rubs were noted at rows #3 and #4 of the turbine.

The knife seals in the row 3 and row 4 inner stage seals exhibited lost material and thermal indications on the seals along with polished shaft surface areas (see Figure 6 below for the seal condition).

Figure 6 – Row 3 Knife Seal Condition

Corrective actions were performed including re-alignment of various components to ensure that shaft/seal clearances were within design criteria during the next assembly.

Reassembly of the machine was performed without indication of any other abnormalities.

<u>After Repair Startup Vibration Analysis:</u>

After reassembly was completed the unit was prepared and restarted to determine the effect of the corrective actions. Figure 7 below shows the initial startup after the corrective actions. The overall high vibration amplitudes had improved by approximately 1 mil to a peak of 8.5 mils.

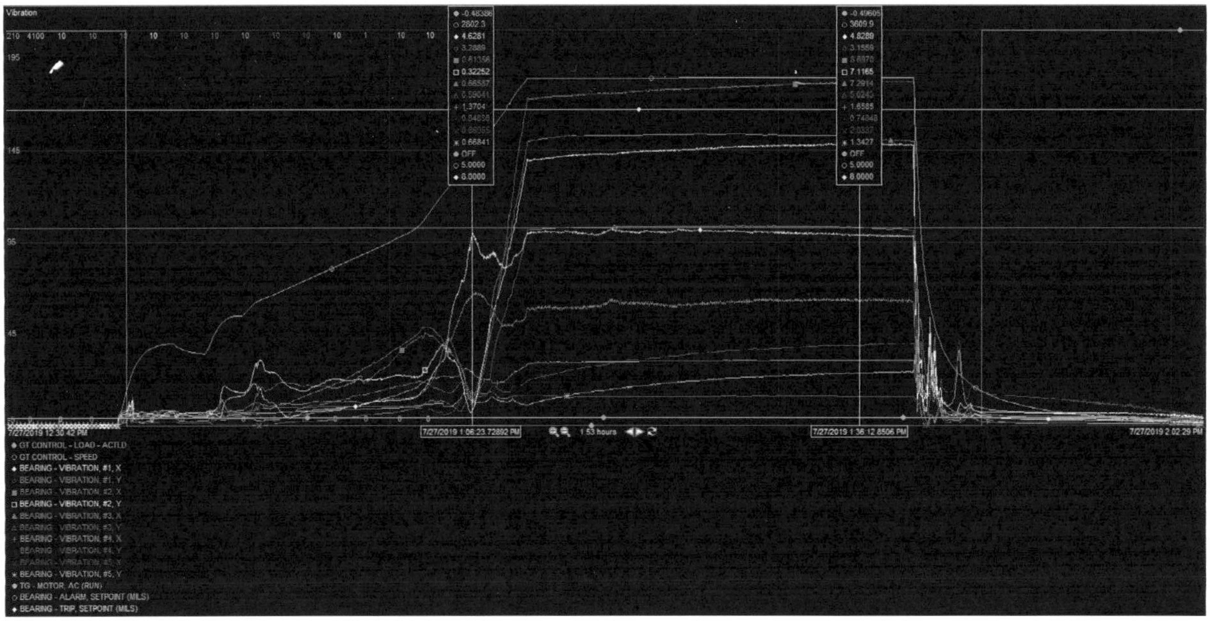

Figure 7 – Startup After Corrective Actions

The plot below (Figure 8) shows the Direct and filtered 1X trends for the initial operation.

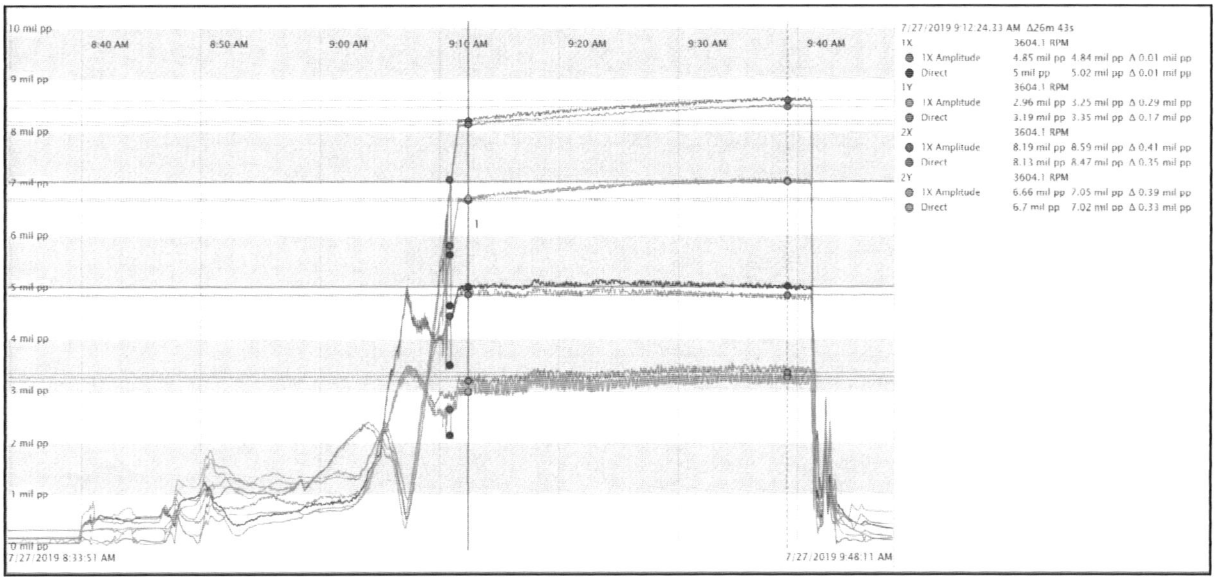

Figure 8 – Direct and 1X Amplitude Trends

Figure 8 indicated that the synchronous energy continued to be the dominant influence in the behavior of the machine with very little difference between the 1X and Direct readings. Figure 9 below shows the Cold Slow Roll (top table), Full Speed No Load (middle table), and Hot Slow Roll (bottom table) conditions from the initial operation.

7/27/2019 8:40:30.38 AM

Name	Speed	Gap	Direct	1X	1X Phase	2X	2X Phase	N	NX	NX Phase
1X	310.9 RPM	-10.88 V	0.48 mil pp	0.30 mil pp	202°	0.18 mil pp	6°	0.5 X	0.01 mil pp	
1Y	310.9 RPM	-10.81 V	0.45 mil pp	0.31 mil pp	286°	0.17 mil pp	185°	0.5 X	0.01 mil pp	
2X	310.9 RPM	-9.28 V	0.43 mil pp	0.13 mil pp	152°	0.25 mil pp	224°	0.5 X	0.01 mil pp	
2Y	310.9 RPM	-9.07 V	0.68 mil pp	0.13 mil pp	194°	0.37 mil pp	57°	0.5 X	0.01 mil pp	

7/27/2019 9:12:14.51 AM

Name	Speed	Gap	Direct	1X	1X Phase	2X	2X Phase	N	NX	NX Phase
1X	3604.1 RPM	-9.01 V	4.97 mil pp	4.86 mil pp	36°	0.29 mil pp	64°	0.5 X	0.08 mil pp	251°
1Y	3604.1 RPM	-10.03 V	3.16 mil pp	2.99 mil pp	101°	0.22 mil pp	208°	0.5 X	0.02 mil pp	95°
2X	3604.1 RPM	-7.72 V	8.1 mil pp	8.19 mil pp	342°	0.30 mil pp	214°	0.5 X	0.02 mil pp	
2Y	3604.1 RPM	-7.60 V	6.69 mil pp	6.68 mil pp	72°	0.21 mil pp	37°	0.5 X	0.04 mil pp	

7/27/2019 9:46:55.11 AM

Name	Speed	Gap	Direct	1X	1X Phase	2X	2X Phase	N	NX	NX Phase
1X	322.3 RPM	-9.78 V	0.37 mil pp	0.21 mil pp	194°	0.16 mil pp	8°	0.5 X	0.00 mil pp	
1Y	322.3 RPM	-11.88 V	0.36 mil pp	0.20 mil pp	282°	0.15 mil pp	185°	0.5 X	0.00 mil pp	
2X	322.3 RPM	-8.93 V	0.44 mil pp	0.13 mil pp	152°	0.24 mil pp	226°	0.5 X	0.00 mil pp	
2Y	322.3 RPM	-8.56 V	0.67 mil pp	0.14 mil pp	204°	0.37 mil pp	58°	0.5 X	0.00 mil pp	

Figure 9 – Tabular List Data

The Tabular List data indicated minimal variations in the Cold/Hot Slow Roll conditions. The Orbit Timebase Direct data as shown in Figure 10 at full speed no load indicated open forward precessed Orbits for both bearings. The #1 bearing was the more elliptical plot indicating a possible preload or asymmetrical stiffness in a generally horizontal direction.

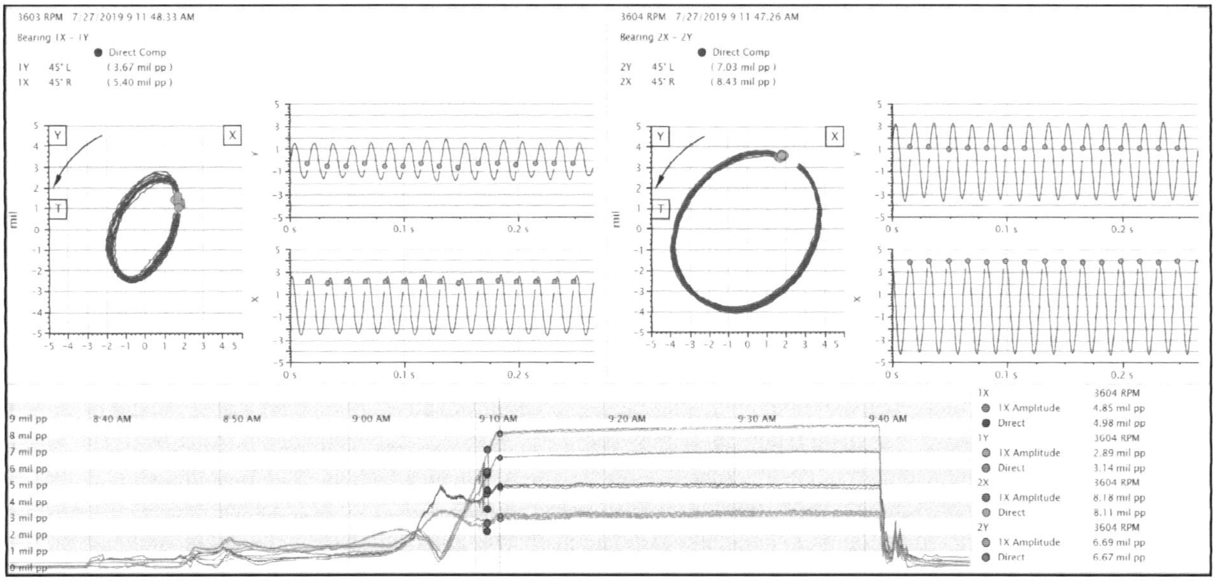

Figure 10 – Orbit Timebase

The Bode plot review as shown in Figure 11 indicated improved (lower) amplitudes during startup (black trace) and steady state (blue trace) for the #1 and #2 bearings. During the coast down (red trace) the response appeared to be shifting to a more typical pattern.

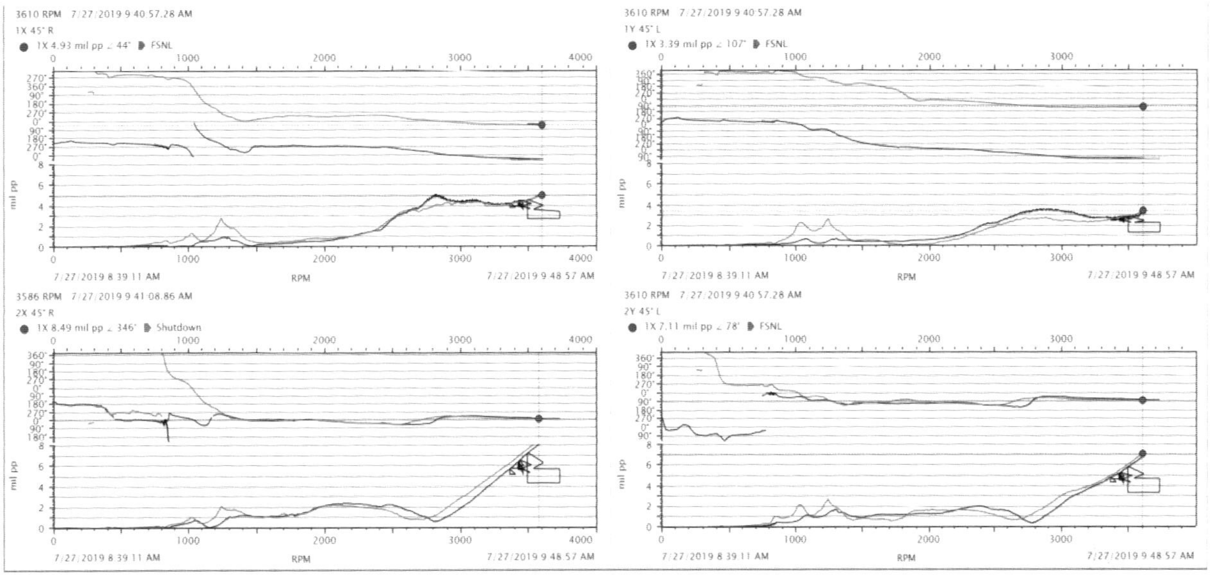

Figure 11 – Bode Plots

The Polar plot review (note that there was an over speed test performed) as shown in Figure 12 indicated similar improvement indications in behavior during the coast down.

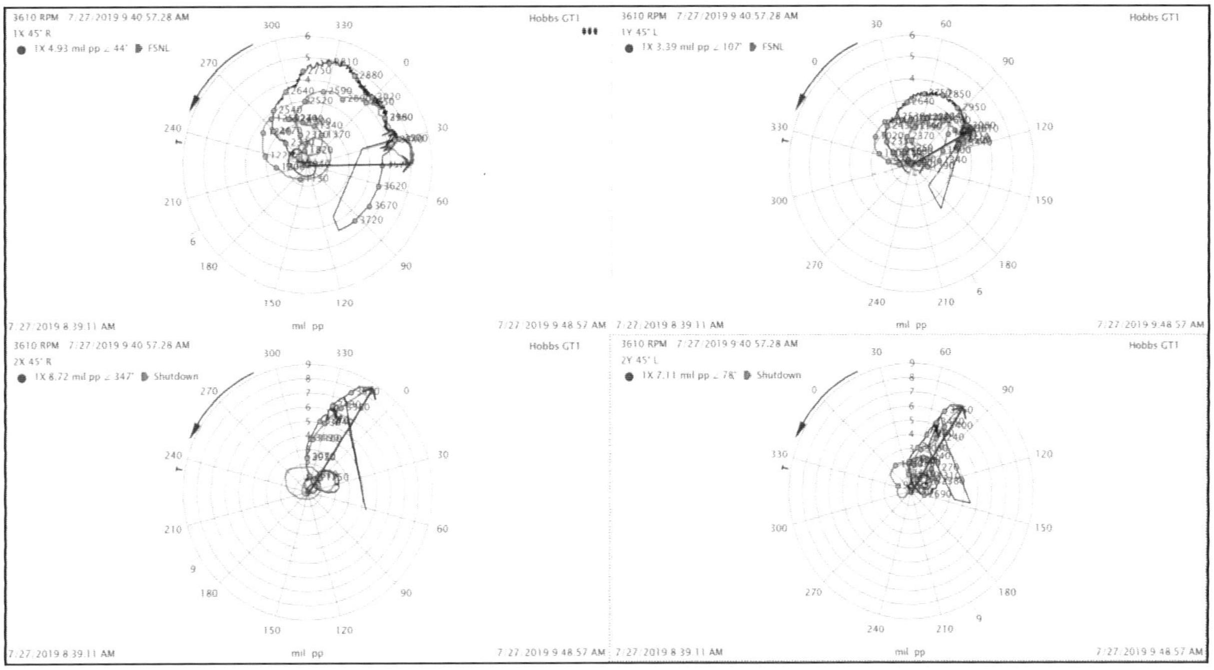

Figure 12 – Polar Plots

The Shaft Centerline plot review as shown in Figure 13 indicated typical gas turbine behavior during a cold startup and short duration run. The #1 bearing exhibited horizontal movement during the thermal growth/stabilization portion of the operation, while the #2 bearing exhibited a vertical thermal growth/stabilization indication. Both bearings dropped between 8 and 10 mils during the coast down.

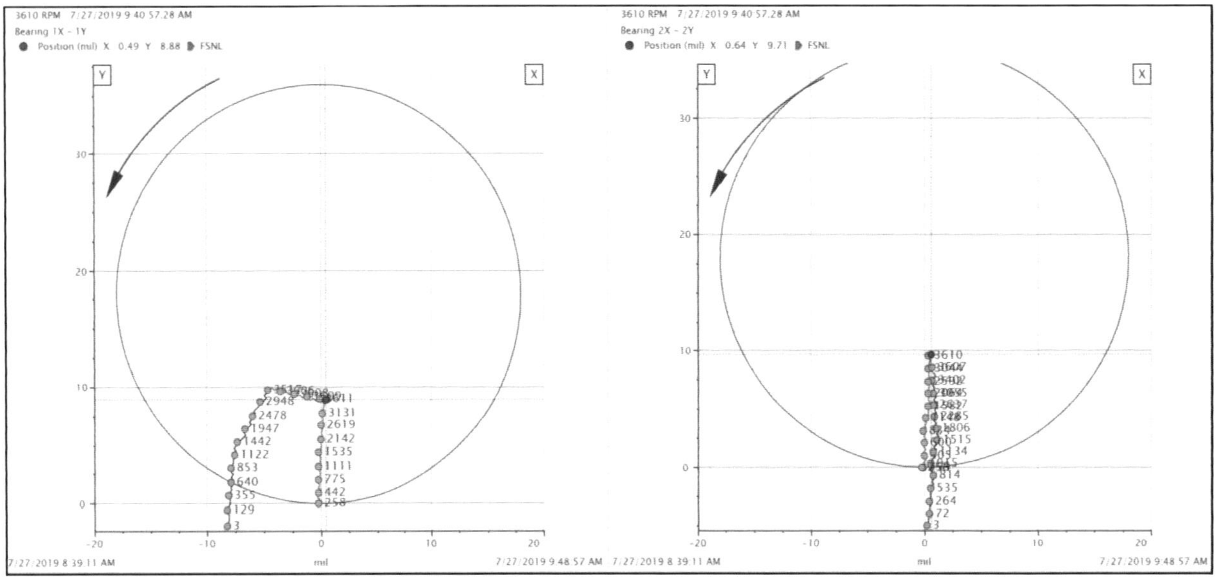

Figure 13 – Shaft Centerline Plots

Given the following:

- Improvement in the overall amplitudes during the operation and the coast down
- No specific indications of rubs
- Confidence in the reassembly conditions

It was recommended to proceed with performing trim balancing of the rotor.

Balancing Efforts:

For field balancing of this rotor, vibration data from the ‘X’ proximity probes was chosen as the primary vector measurement to base balance weight mass and phase response calculations from. For the exhaust end, the ‘X’ direction amplitudes were 52% higher than the ‘Y’ direction measurements. For the inlet end, the ‘X’ measurement direction was 23% higher in amplitude. These amplitude directional differences are not unusual and are reflected in the orientation of the major and minor system stiffness axes as shown below in Figure 14 of the filtered 1X orbits from the unit at full speed, no load conditions prior to rotor balancing. Asymmetrical rotor support stiffness is typical and desirable as it indicates proper journal and bearing loading.

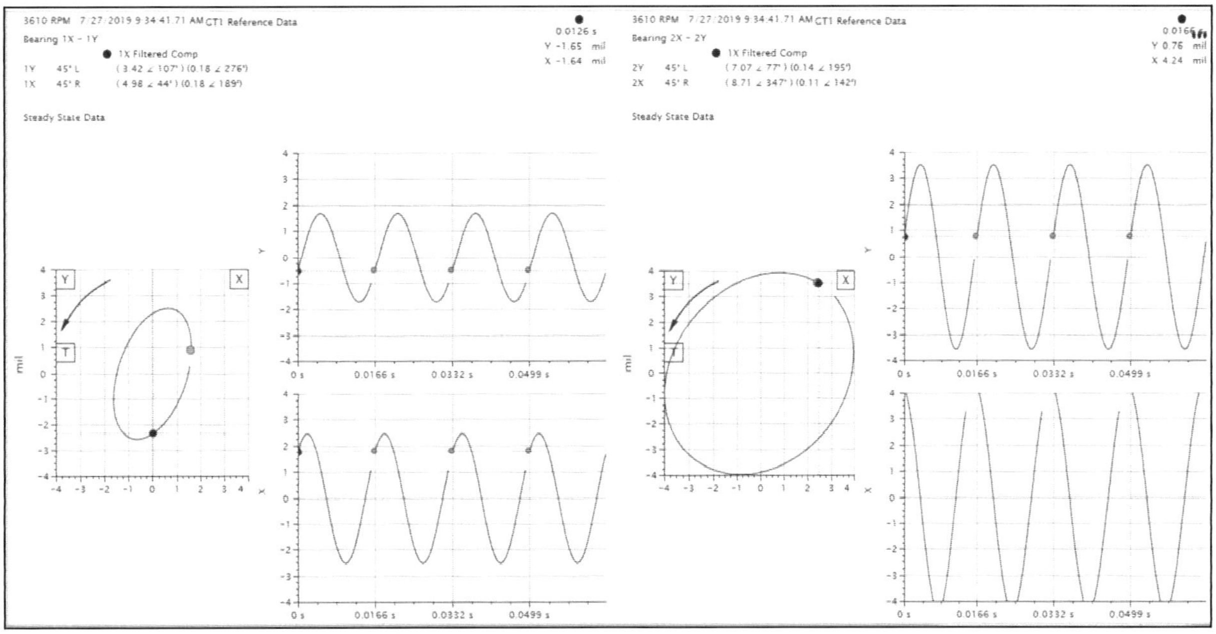

Figure 14. 1X filtered and compensated orbit plots from exhaust (left) and inlet (right) bearings prior to unit trim balancing.

Although the 'X' vibration vectors were chosen as the initial trim weight calculation basis, data from the 'Y' axis probes was also included in all calculations to verify and monitor the rotor response to trim weight additions. The amplitude and mass response to trim balance weights should be proportional in both measured directions for simple rotor unbalance issues and radical departures from this, such as a large response in one measurement direction with little response in the other would be an indication of other mechanical issues that warrant investigation.

Once the primary measurement direction was chosen for trim balance calculations, a rotor balance plane weight map was oriented to the 'X' probe directions for both the inlet and exhaust balance planes as shown in Figures 15 and 16 below.

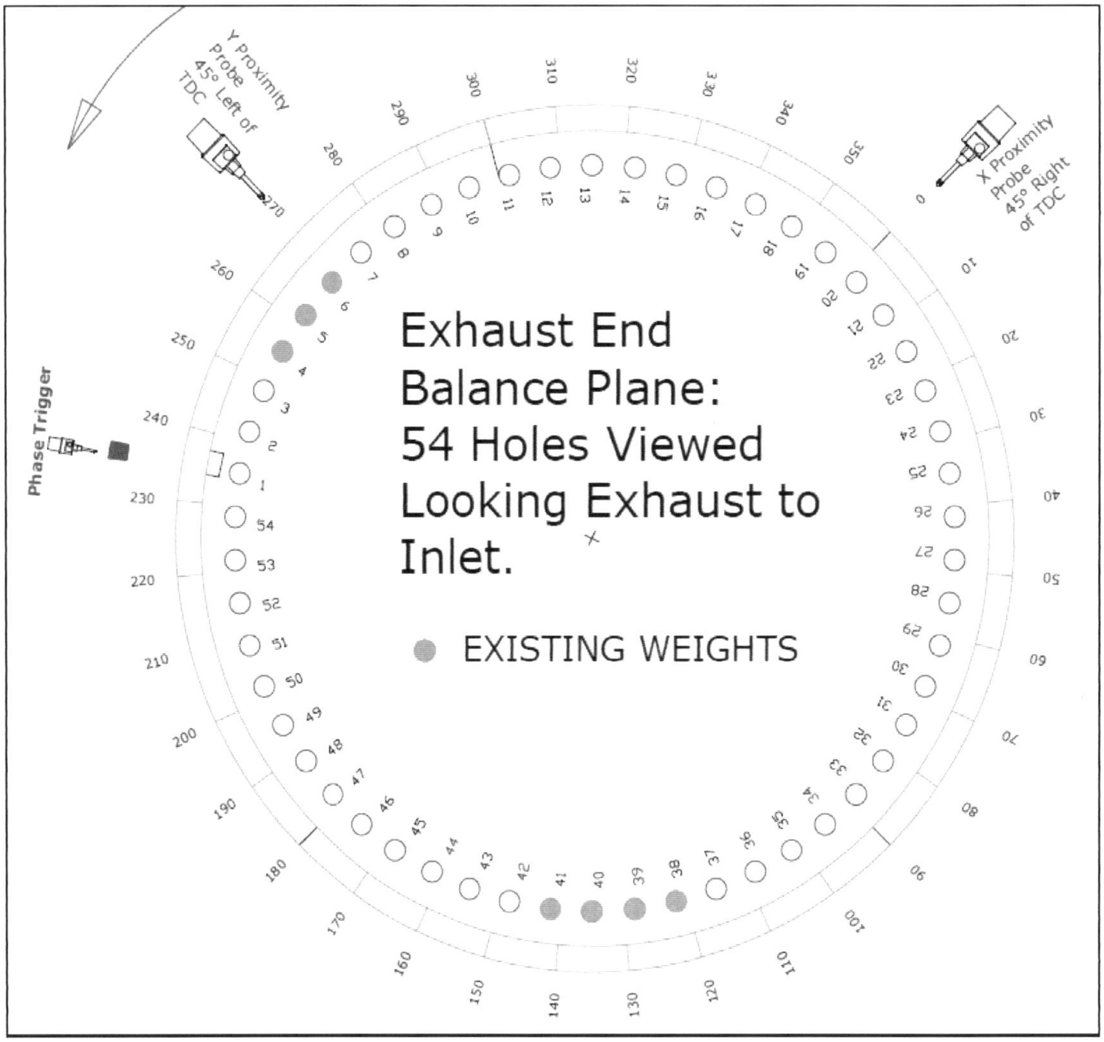

Figure 15. Balance plane map of the exhaust end of the rotor.

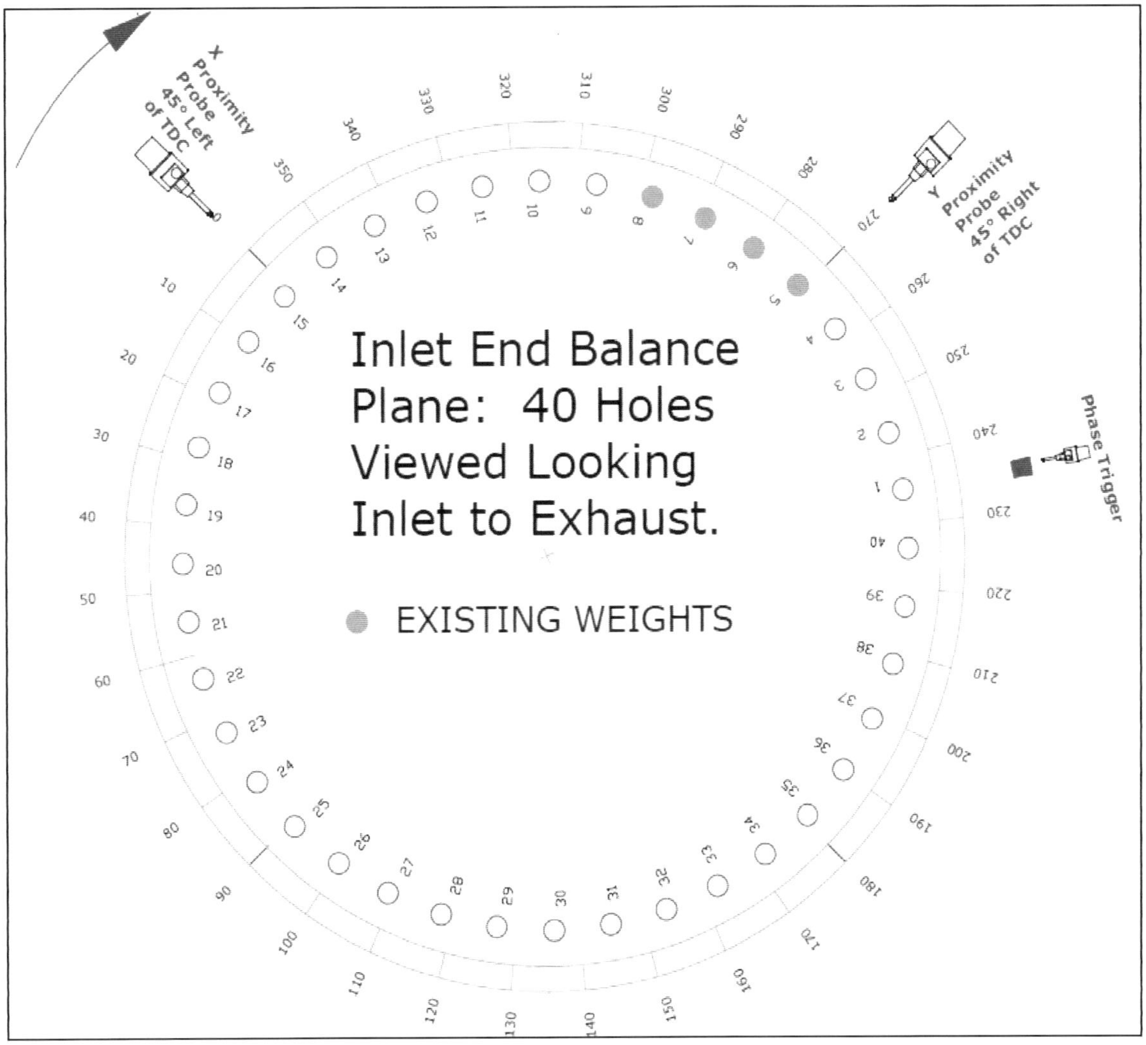

Figure 16. Balance plane map of the rotor inlet end.

The balance weight planes for this rotor are axial direction tapped holes that will accommodate up to a 5.9 ounce weight each. The holes are numbered against rotor rotation direction with hole #1 positioned in-line with the phase trigger notch on the rotor. On the weight map, the phase trigger notch (and hole #1) is located at the angular position of the phase trigger probe. Additionally, vibration phase angles are also displayed starting at zero degrees of phase at the 'X' probe location since the 'X' probes was used for primary calculations. This map arrangement then correctly orients the weight hole positions with the probe phase angles and simplifies the job of locating the correct weight hole relative to the desired phase angle of the calculated weight position. For example, should vibration data and balance calculations call for a desired weight at 190° of phase on the inlet end of the rotor, from the weight map, it's easy to see that would correspond with a weight installed at hole #36. Additionally, on the weight map are displayed existing weights that have been previously installed in the rotor. This allows for adjustment in weight vector calculations should balance calculations indicate the need for weight where existing weight are already installed. It may also allow for the removal of existing weights should calculations call for trim weight installation opposite of existing weights; although, extreme caution is always urged regarding the removal of weights on flexible rotor systems as these weights could be modal balance weights and weight consolidation can lead to undesired results.

On the initial run of the unit, the 1X filtered and compensated (for slow roll runout) vibration amplitudes on the rotor indicated a substantial level of rotor unbalance on the inlet end at over 3800 ounce-inches (oz-in). The exhaust end unbalance was also high and amounted to just under 2000 oz-in. Balance calculations for the initial rotor conditions indicated that although the static (in-phase unbalance end-to-end) amount of unbalance was over 6 mils, there was still a substantial amount of couple (180° out of phase) unbalance on the unit at just under 4 mils and a multi-plane balance methodology was chosen to address the issue.

Traditionally in multi-plane field balancing, calibration weights are installed in each balance weight plane and the rotor is operated to full speed. The data from each of these operational runs is then used in a matrix calculation (generally a least squared solution) to determine an idealized correction weight in each weight plane for a final trim balance. For a rotor with two balance planes, this would require a minimum of 4 runs of the unit assuming that the location and mass of the calibration weight is conservative enough to allow the unit to reach full speed without a trip on excessive vibration and aggressive enough to produce a change in the vibration vectors that is the confirmed result of the balance weight and not typical run to run thermal variation or the result of other machinery non-linear effects such as rubbing. Perfectly, the calibration weights on each end are calculated to reduce vibration amplitudes at each run such that they are not removed between calibration runs and final correction weights are not required.

In addition to the number of operational runs required, the limitation of multi-plane calibration weight balancing lies in the recovery of calculations to any non-linear response to the calibration weights. Realization of any non-linear response to the calibration weight runs can't be determined until the correction weight run (run 4) when the influence of the correction weights can differ from the calculated influence of the calibration weights. If multi-plane correction weights have been installed on what was supposed to be the final run of the unit and non-linear results affected either calibration weight run or the final correction weight run, a pure mathematical recovery of the matrix calculation for a second modified correction weight is doomed because it can't be determined which calibration weight or if the final correction weight produced a non-linear response. Depending on the severity of the non-linear response, recovery may involve a repeatability run of each of the calibration weights to determine if the influence matched previous runs.

Had a static/couple balancing methodology been chosen, since the rotor had both a significant amount of static and couple unbalance, a static calibration weight run would have been followed by a static correction weight run which would then have been followed by a couple calibration run and then a couple correction weight run, also for a total of 4 runs. The difference being the additional time and man power required to install calibration and corrections weights in each end of the rotor which would have also required additional cooling time for entry into the hot end of the rotor for the exhaust end weight installations.

In order to minimize project time and operational runs for trim balancing the unit, the first single plane weight in the unit was installed in the exhaust end of the rotor to reduce inlet end amplitudes and exhaust end amplitudes as well. Most large frame gas turbines respond similarly to this opposite end trim balance approach. Typically, the in-plane amplitude response to balance weights is on the order of half or less to the opposite plane response to the weight. For these models of gas turbines, the phase of the response to balance weights is similar on each end of the rotor and although the amplitude response is largely different, since the unbalance vibration vectors were less than 90° apart, calculations indicated positive response to the weight on the inlet end and the exhaust end.

Although no individual influence data was available for this unit, family influence response data generally provides very good results and is based on the fact that similar units produce a close cluster of influence responses to balance weights. A typical plot of balance weight influence response data is shown below in Figure 17. This polar plot shows the amplitude/unit mass and phase response to balance weights for a family of different gas turbines of the same or similar make and model. Balance influence response data for a unit is invaluable as it allows for 'one-shot' correction weights to be installed assuming a largely linear response. An average of the influence family of amplitude/unit mass and phase location response was used for the exhaust end weight calculation. The 3800 ounce-inches of unbalance on the inlet end called for an aggressive move of several full size balance weights in the exhaust end.

The calculated correction mass was over 60 ounces for a full reduction of the 8+ mils of amplitude on the inlet end; however, these very large mass results are typical for high amplitude balance weight calculations and are generally never installed for several reasons. The first reason for not installing the exact calculated mass amount is that we are never attempting to balance a rotor to zero vibration amplitude. In most all industrial machinery, vibration amplitudes at operational speed that are below ISO 20816-2 Zone A/B levels for new or re-commissioned units is more than acceptable for safe, reliable operation. For this unit, anything at or below 3.1 mils was acceptable. The second reason for not installing the full calculated balance weight amount is that generally, as vibration amplitude decreases, the rotors response to weight can become more sensitive as issues like light seal rubbing disappear with decreased vibration amplitudes. Another reason for not installing the full calculated mass amount is the diminished effect of that weight as the weights are spread over a large arc of the balance weight plane. Lastly, and perhaps most importantly, is that although there is good reason to believe that the rotor will respond linearly and similar to the family of influence data used to calculate the balance weight, without a calibration weight run on the rotor, the true response of the rotor to balance weight is unknown and some conservativeness in the weight installation is prudent.

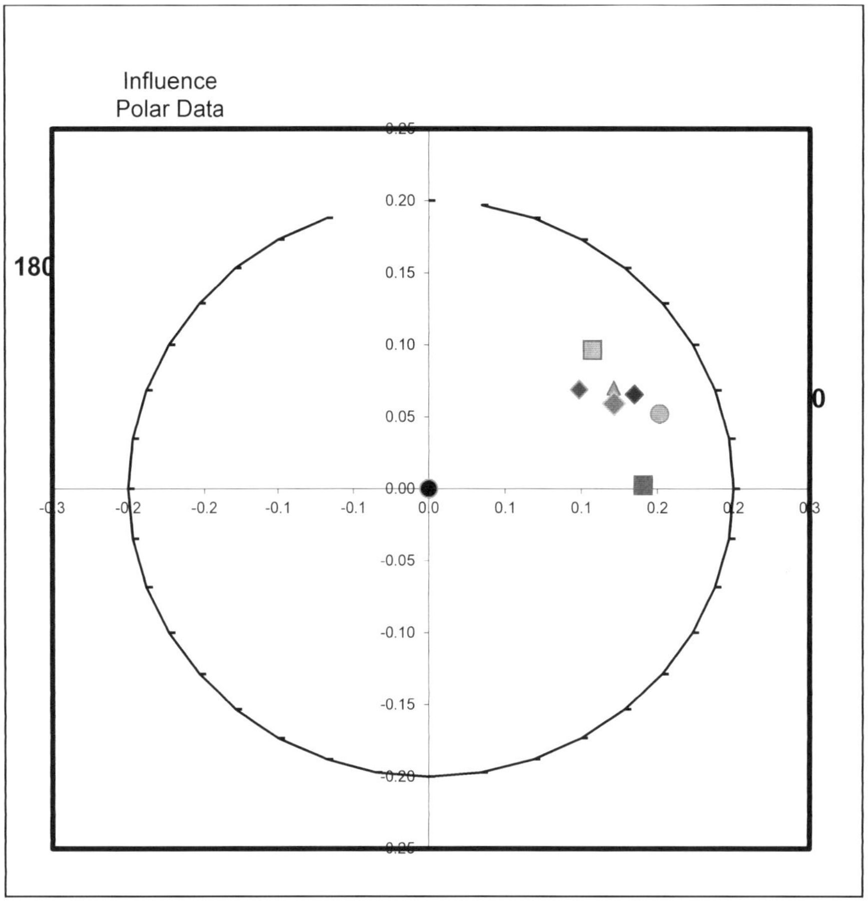

Figure 17. Polar plot of balance influence response.

The exhaust end weight that was installed was 7 full size weights with a total mass vector summation effect of 38.4 ounces. The unit was operated to full speed and displayed a significant reduction in direct and 1X amplitudes. Figure 18 below shows the 1X filtered and compensated orbits after the exhaust end weight installation. The 8.6 mils ∠ 346°at the inlet end was reduced to 4.61 mils ∠ 348°; with the exhaust end dropping from 4.93 mils ∠ 44° to 3.64 mils ∠ 16°. Although these operational amplitudes are acceptable for unrestricted operation per ISO 20816-2, amplitudes at or below 3.1 mils were desired and the trim balancing program was continued.

The next weight installation was calculated for the inlet end of the unit. Although the effect of inlet weights on the inlet end are not as great as exhaust end weights (~only 43% mass effect), they do still have inlet end effect and are much faster to install because the exhaust end bearing tunnel does not need to be extensively cooled for human habitation and they are easier to install because of balance plane accessibility. The inlet end balance weight was also calculated from the family of influence data available and was 4 full weights for a vector mass effect of 24.5 ounces. The results of this weight is shown below in Figure 19. The final inlet end 1X amplitudes were 1.69 mils with exhaust end amplitudes at 0.76 mils in the 'X' directions.

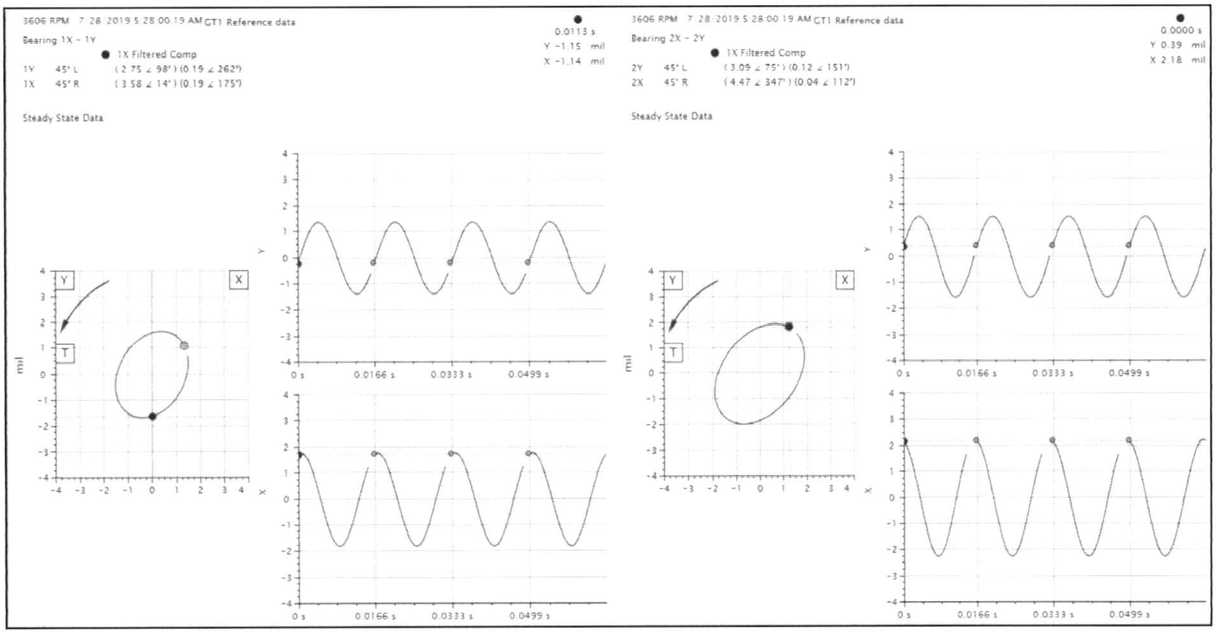

Figure 18. 1X compensated orbits after the turbine exhaust end balance weight.

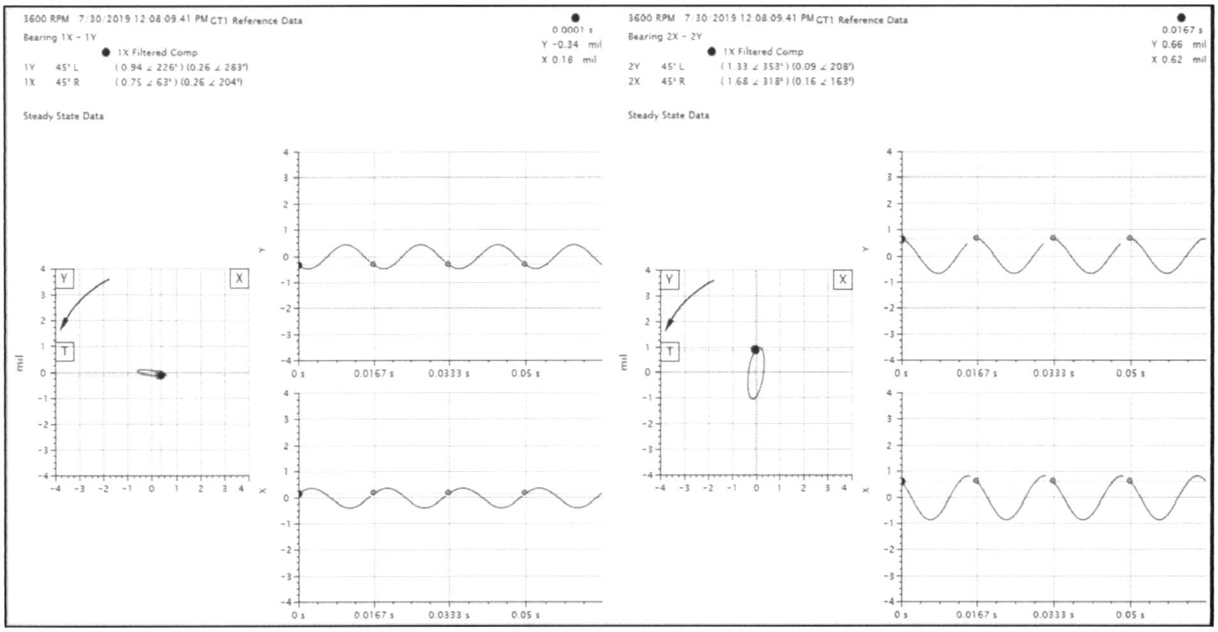

Figure 19. 1X compensated orbits after the turbine inlet end balance weight installation.

It should be noted that even though the filtered and compensated 1X Orbits were highly elliptical, at the amplitudes observed (both transient and steady state) and obtaining the expected response from the balance weights, we did not consider this Orbit shape to be a concern. It is likely that additional influences were driving the apparent preload, but it did not appear to be detrimental to present machine operation and we did note it for future reference. For gas turbines with soft bearing support stiffness, bearing seismic amplitudes can be a factor affecting the orbit shape and amplitude. Bearing seismic data was not available on this unit for determination of the shaft absolute amplitudes.

It was also noted that after ~ 6 months of operation the bearing 1Y the overall amplitude had changed ~ 0.5 mils, the other 3 channels had changed ~ 0.3 mils or less.

Conclusion:

This case history is a very good example of the very common 'Chicken or the Egg' syndrome in machinery vibration analysis. Is the rotor primarily out of balance and causing rubbing or; is the machine rubbing (from tight seals or other issues) and causing rotor bowing that creates unbalance (and often harder rubbing)? This issue is often more confused by the typical first solution tried: attempt to trim balance the rotor.

If the rotor is severely out of balance and rubbing, the rubbing generally bows the rotor further creating more imbalance and because of the rubbing, the rotor response to trim balance weights is usually non-linear or atypical from known influence response creating confusion about the root cause of the issue.

If the rotor is rubbing and causing rotor bowing and imbalance, the results of trim balancing will be similarly non-linear. If trim balancing is moderately successful, as seal rubbing clears with operation time, the rotor has now actually been unbalanced with the addition of the initial trim balance weights and will gradually increase in amplitude as the rubs begin to clear.

For these reasons, it is important to have a good understanding of the work that has been performed on the unit to determine if rubbing from tight (seal) clearances is likely to be the root cause of the issue, or if blade or blade row replacement could be creating imbalance and the root cause of the issues. Further, being able to determine the telltale signs of rubbing vs imbalance in the vibration data such as increasing amplitude with speed with little phase changes; high coast down slow roll amplitudes from rotor bowing; flattened or pre-loaded orbits; sharp directional changes in orbits; distorted shaft centerline positions and abnormal frequency content will help in determining proper mitigation and corrective steps needed. Often, not rushing to install balance weights and continued operation of the unit or repeated cycling may begin to reduce or even clear seal or packing rubs that will determine if rotor trim balancing is required. Lastly, borescope inspection may be required to determine root cause and corrective steps, including rotor trim balancing.

For the final trim balancing of this unit after correction of the seal rubbing problems, the multi-plane balancing methodology based on family influence data available combined with some experienced judgement on the machine behavior reduced the number of operational runs required to trim balance the unit from 4 runs to 3 runs and the project time by an estimated 3 to 4 days.

The root cause of the large unbalance on the rotor was never determined. Spot checks of the blade numbering during the disassembly of the unit to correct the seal rubbing issues indicated blade rows to be installed per the mass-moment sequencing calculated for each of the blade rows. A comprehensive review of the mass-moment sequencing calculations for each of the blade rows did not indicate any errors of calculation. Rotor runouts taken during the seal clearance outage did not indicate any changes in the rotor runout conditions from prior to the outage. One possible root cause may have been errors in the individual blade mass-moment measurements or possibly instrumentation calibration errors during the weighing of blades.

NATURAL FREQUENCY TESTING AND VARIOUS TECHNIQUES

Author: Thomas. B. Lantrip, P.E.

NATURAL FREQUENCY TESTING AND VARIOUS TECHNIQUES

Thomas. B. Lantrip, P.E.

Consulting Engineer - Southern Company Services
Fleet Services – Performance and Reliability
3535 Colonnade Pkwy
Birmingham, AL 35243

tblantri@southernco.com

Abstract: Often the vibration concern for equipment or components is due to the existence of closely spaced natural frequencies to the excitation frequency. This resonance condition requires the identification of the natural frequency, and possibly the mode shape to associate the coupling or effect on the vibration amplification. Natural frequency testing may include several techniques to capture the required natural frequency data. Several obstacles can be encountered in pursuit of this natural frequency and mode shape determination, including the size of the component or structure, adequate excitation methods, correct instrumentation, time, and budget constraints. This paper discusses the various methods, insight into trials and tricks, some failed some successful to identify the natural frequencies of interest.

Keywords: Natural frequency; impact testing; mode Shapes; forced Response; ambient response.

Vibration Analysis Consideration:

As part of the vibration analysis, it is often important to understand or determine the relationship between the forces associated with machine of interest and the response characteristics. For example, we often find that the vibration measurement on a particular machine or structure typically has characteristics that are preconceived before we start collecting the vibration response measurement. Case in point, the horizontal centrifugal machine often has a much higher response in the horizontal direction (perpendicular to the shaft axis) as compared to the vertical direction (up and down). As we know this condition is not an absolute in every case, but for a broad cross-section of machines it appears to be statistically expected in most cases. Also, in keeping with same horizontal machine, we expect axial or longitudinal (down the rotational axis) vibration response to be expected to be much lower than the horizontal (transverse) direction and most cases the vertical direction.

Figure 1 : Motor Driven Boiler Feed Pump Train

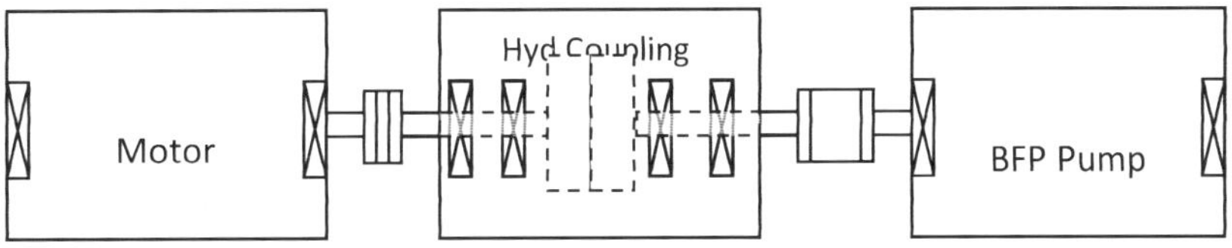

Natural Frequency Determination:

As we break down the vibration problem into decision points, excessive or unusual levels of vibration can lead us to use the preceding screening filter to identify our condition and start to form questions about the character of the concerning vibration problem.

Whenever we encounter a condition that indicates a higher vibration amplitude, we would possibly look to the frequency spectrum to perform a review of the spectra. Based on the spectral analysis we may question the shape and location of the harmful frequencies. This analysis could also result in questioning the relative proportions of the vibration in the 3-dimensional space of the measurement location. An example of a spectral content that may raise some questions. In this case the vibration amplitude in one direction (perpendicular) is 3.5 times higher than the other direction (parallel to discharge).

Figure 2 : Vertical Pump with Vibration Spectra at Top of Motor (in/s pk)

As we know from the academics, the natural frequency can and often will amplify the existing forced vibration due to the interaction between the natural frequency and normal vibration response. When this condition exists where the exciting force frequencies and the natural frequency of the system are in similar proximity, we call this resonance. The response of the machine is being amplified due to the natural frequency.

Generally, we could lead ourselves to question whether the natural frequency of the machine or structure was being naturally amplified. With this we would possibly need to use a testing technique to determine the natural frequency of the machine/structure at the harmful range of frequency.

Natural frequencies can be identified in many ways. The most straightforward and technique that provides the highest-quality results is the forced impact using an instrumentation force-cell hammer and a response sensor like an accelerometer or velometer. Strain gages or displacement probes can be used also.

Figure 3 : Impact Hammer and Various Seismic Sensors

(Accelerometers and Velometers and Medium and Large Impact Hammer)

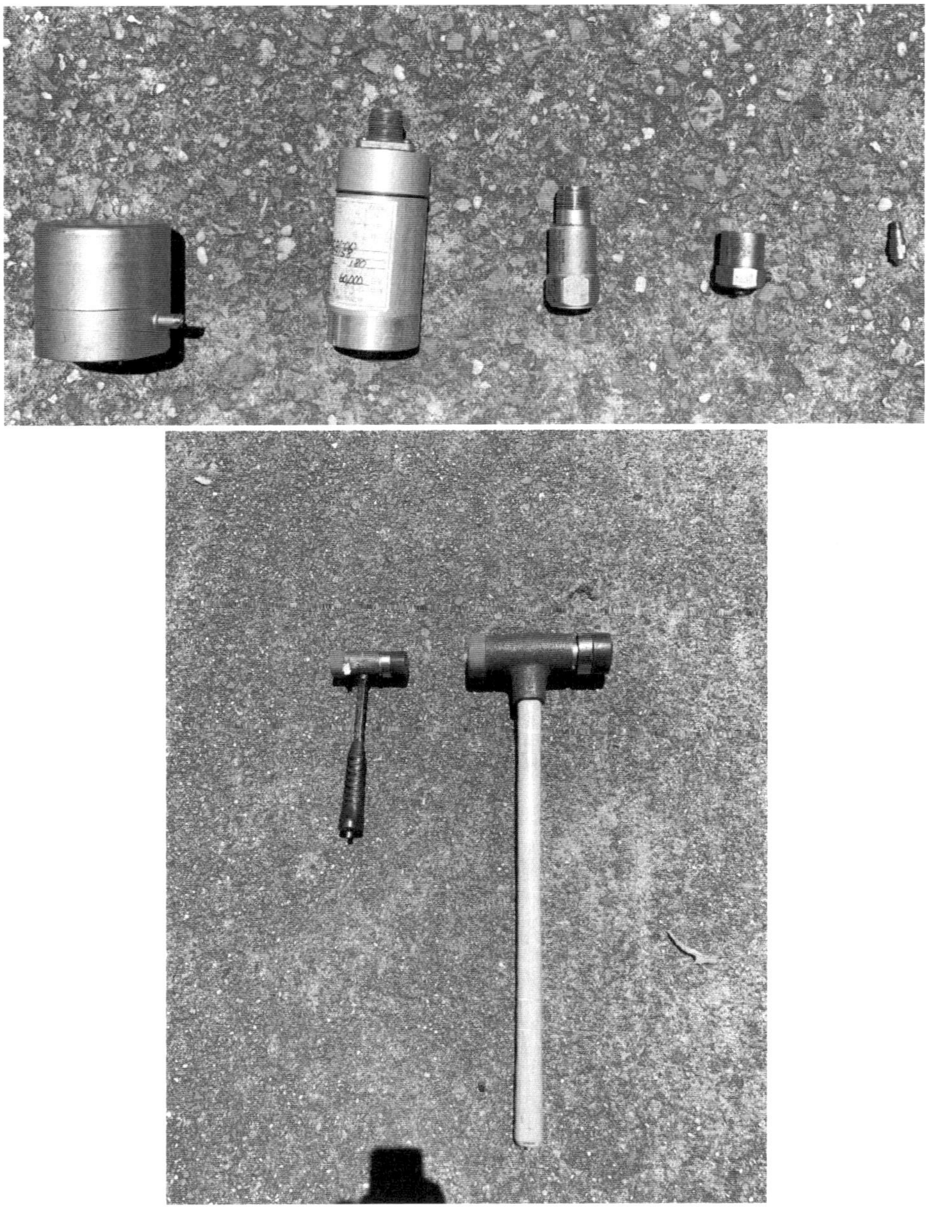

Figure 4 below provides the basic time history plot of the force from the hammer and the response of the accelerometer below. Windows are applied to the force and acceleration waveforms and averaging techniques are transformed using the Fast Fourier Technique (FFT) to arrive at the plots in Figure 5.

Figure 4 : Impact Time History and Response

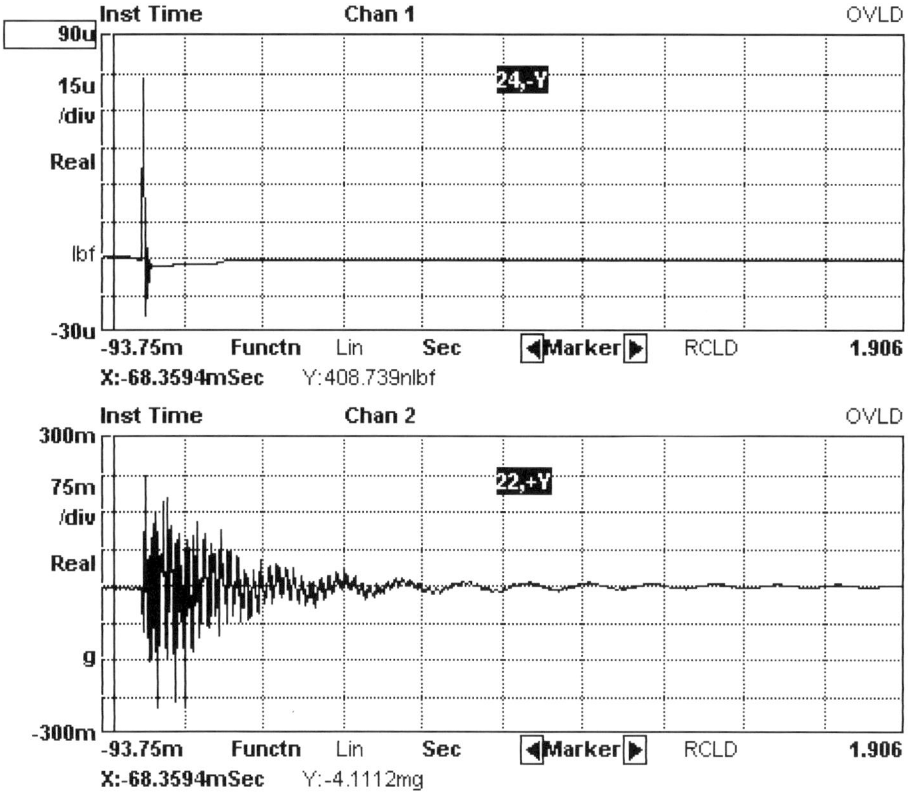

Figure 5 provides the Frequency Response Function (FRF) or Transfer Function (TF) between the response of the accelerometer and the force hammer. The FRF provides the spectral content of the response from the impact testing, where the amplified peaks (log scale) identify the natural frequencies. In addition, the phase component of the response is provided in the upper portion of the FRF or TF plot. Normally, where the phase shifts significantly and the amplitude peaks in the spectrum locates the singular natural frequency. Structures and machines have an infinite number of natural frequencies, however thankfully most of the ones that we are interested in are in the lower frequency portion of the spectrum where the energy and response is much more apparent.

Figure 6 and 7 provide the layout of the tested structure, in this case it is a concrete utility pole configured in a horizontal pendulum to identify structural characteristics of the component itself. This example is somewhat an academic case, where the boundary conditions are simplified with the pendulum condition. All the testing involved the lateral impact which eliminated the effect of the pendulum cables, when comparing to analytical results. Factory and manufacturing testing often utilize this approach to normalize and control the known conditions. In-situ testing creates major obstacles to modal testing, as the boundary conditions may be different in many cases, that often challenges the manufacturing designer or owner when problems occur.

Figure 5 : Frequency Response Acceleration/Force Transfer Function

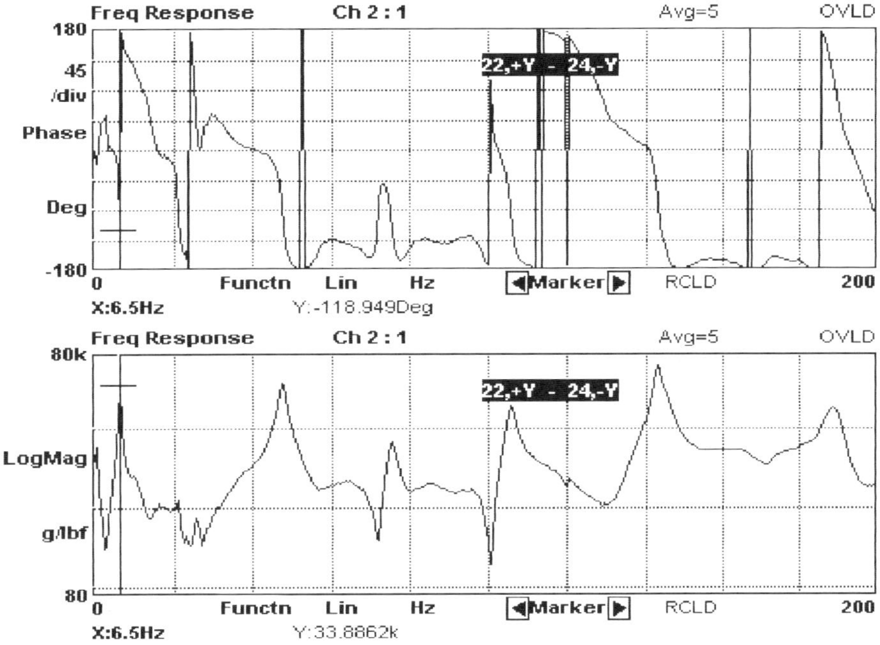

Figure 6 : Tested Structure Cross-Section Description and Layout

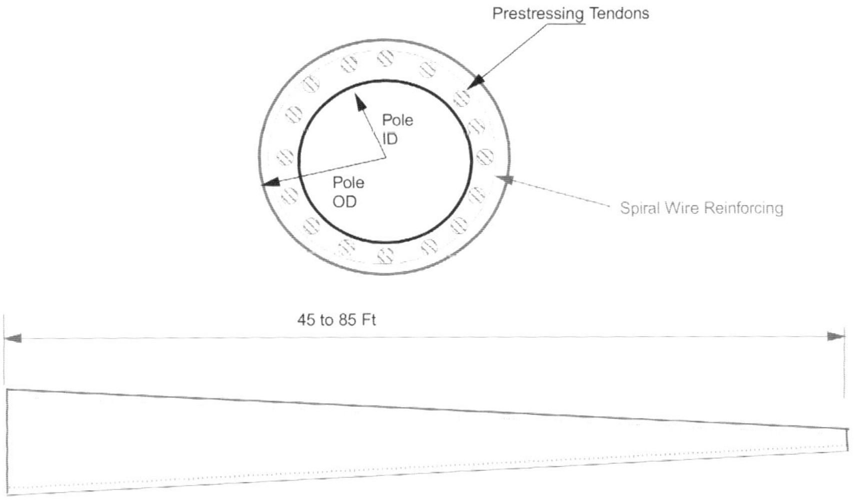

Figure 7 : Tested Component Elevation and Layout

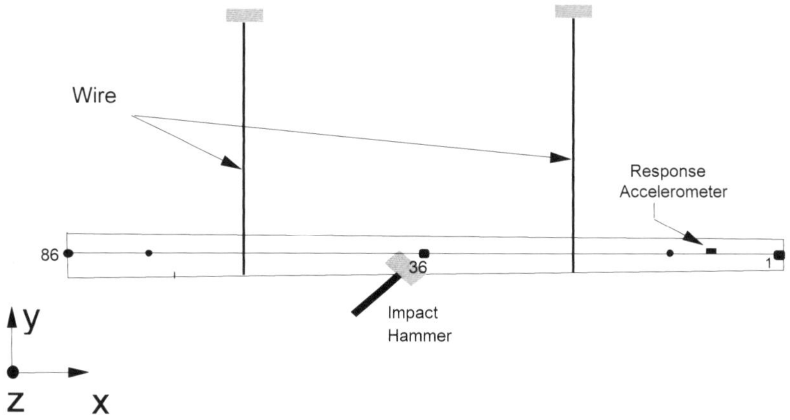

Mode Shapes are developed by comparing the dynamic responses at various locations along the structure or machine and compiling the relative differences in amplitude and phase for each natural frequency. This type of analysis and compilation is usually referred to as Modal testing. Modal testing can be accomplished at the natural frequency to create the natural mode shapes. Operating surveys are akin to the modal survey methods, but are primarily associated with running speed frequencies.

Vertical Pump Natural Frequencies:

Vertical pumps in my experience tend to have the most common condition associated with resonance. The major issue stems from these categories based on my experience.

1. Incorrect Assumptions used for Boundary Support Conditions
2. Inaccurate Modelling Methods and lack of Correlation to Actual Test Data
3. Lower Submerged Column and Pump Components Difficult to Measure and Acquire Test Data
4. Lack of good communication between the vendor supplier and owner for the foundation requirements.
5. Changes to the support stiffness and damping due to degradation of the bolting interface to concrete or changes to the discharge piping support.

Most of the time, the vibration and degradation may be coming from the impellor of the pump deeply submerged in a water or slurry basin. Without easy access, difficulty exists to adequately monitor and quantify vibration and conditions.

Vertical pumps can be quite large and difficult to impact with a normal instrumented sledge-hammer and obtain proper results. Another method that has shown to be quite effective is the use of the ambient in-situ method.

Ambient Natural Frequency:

Figure 8 : Vertical Pump Natural Frequency Determination using Ambient Linear Averaging (Adjoining Pump Operating)

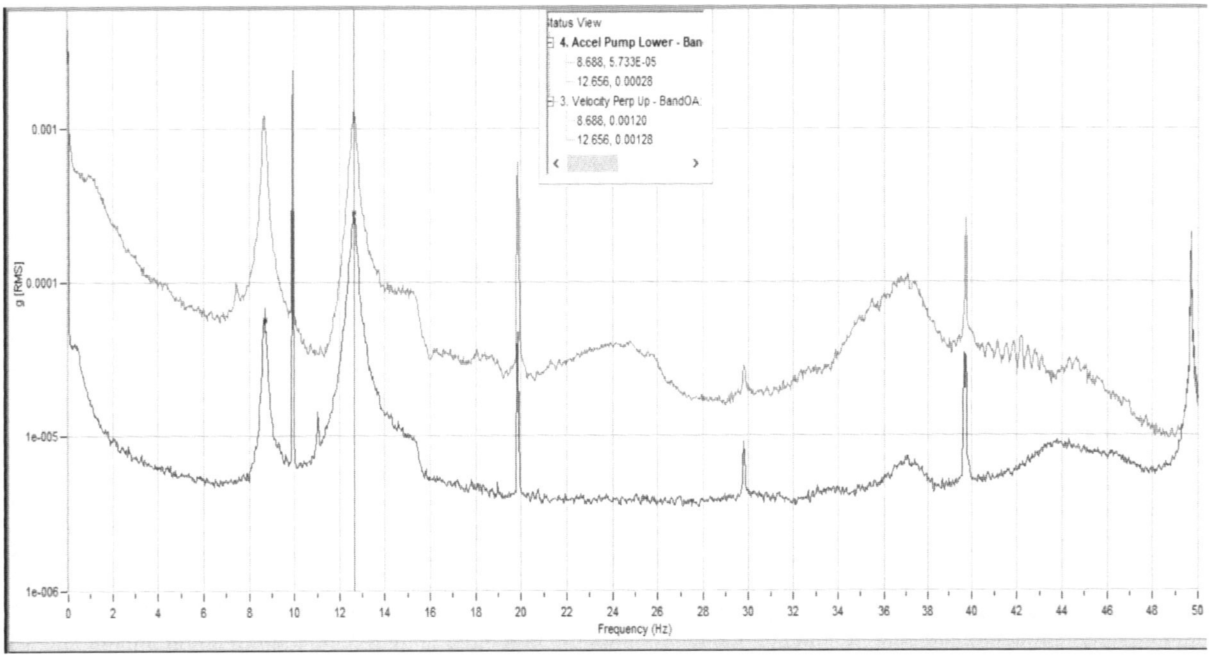

Figure 8 provides an ambient frequency spectrum with the vertical pump not operating using a technique that has the spectra averaged by allowing 200 averages for various points on the top and bottom of the pump column. Typically, vibration has been measured on the top of a vertical pump motor, but with pump degradation issues at the bottom of the pump at the impellor there has been interest in measurement of the bottom of the pump column. Review of the data in Figure 8 shows an apparent natural frequency of 8.688 hz and 12.656 hz, where the operating speed of 9.936 hz. Based on the spectral plot, one can see that lower frequency at 8.688 hz is likely creating an amplification of the operating speed.

Obviously, the impact of the lower column is difficult or impractical. The top of the motor and pump can be practically impacted with a force hammer but without ladders and/or scaffolding this is difficult also.

Figure 9 : Pump Upper Column and Motor and Pump Lower Column below Basin Concrete Base Support

Forced Response:

Using the forces internal to the machine to identify the natural frequencies is a common technique if the machine speed can be varied. Variable speed machines are very good to perform this technique. This technique is one of the best because it utilizes the inherent forces (unbalance normally) and allows easily obtainable information. This data provides the example of the natural frequency (critical speed of a rotor element) of a steam turbine as it is coasted down from the nominal operating speed of 3600 rpm as shown in Figure 10.

Figure 10 : Natural Frequency Response using Bode' Plot

(Amplitude and Phase) from Coast-down on Steam Turbine

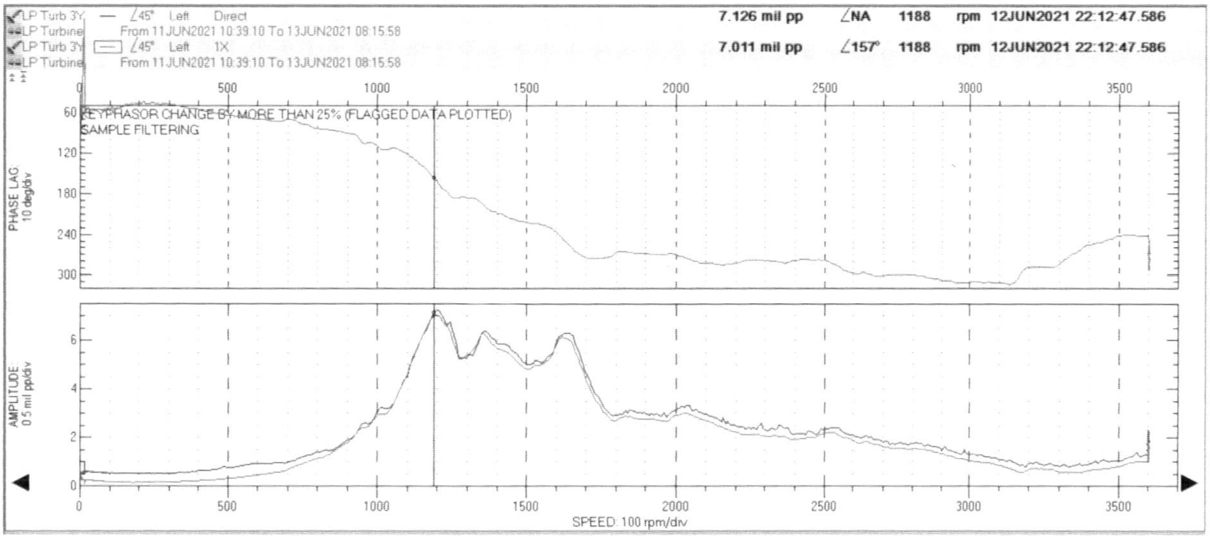

The vibration data shows the maximum response at 1188 rpm which also indicates a significant phase shift at the same speed. This indicates a critical speed of the Low Pressure (LP) turbine rotor. This is one of most preferred methods for obtaining critical speed natural frequencies with the fluid-film bearings. In some turbine rotor trains involving multiple large utility rotors there are also complete train frequencies that have carryover response from adjacent rotor spans.

Figure 11 : Natural Frequency Response using Polar Plot

(Amplitude and Phase) from Coast-down on Steam Turbine

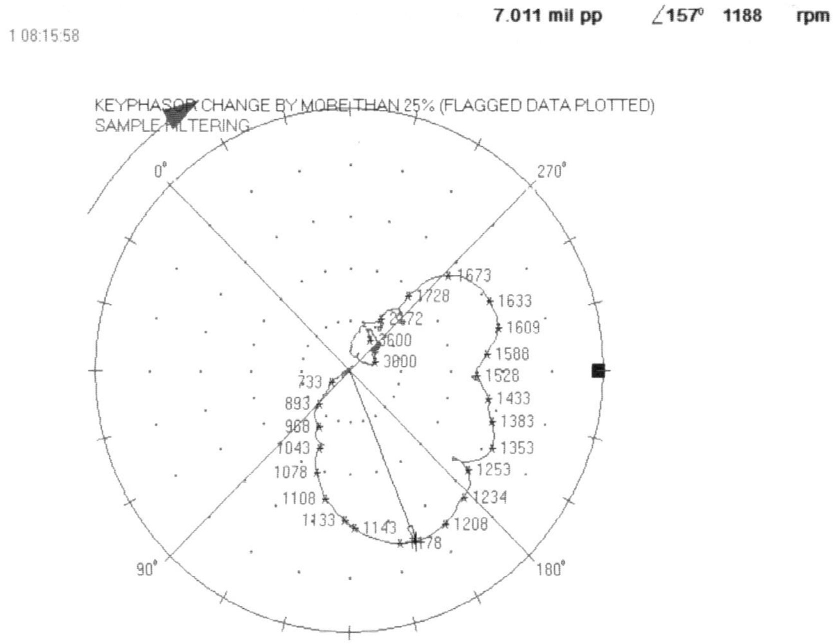

Figure 11 is the polar plot of the same data shown in Figure 10 which indicates the same conclusion of the critical speed at 1188 rpm.

Another case where the natural frequency is not identified below the normal operating speed, but some speculation exists that the natural frequency is just above the operating speed. Figure 12 below provides for a coast-down of an Induced Draft (ID) fan. The peak response of the Bode' plot indicates the highest response just at the top of the operating speed of 900 rpm nominal. Some doubt that this is the natural frequency is observed at the small change in phase angle in the upper portion of the plot. Figure 12 below indicates the 1X order or the operating speed frequency as a function of the rotor speed.

Figure 12 : 1X Order Tracking Bode Plot (Vibration and Phase vs Speed)

Nominal Operating Speed = 900 rpm

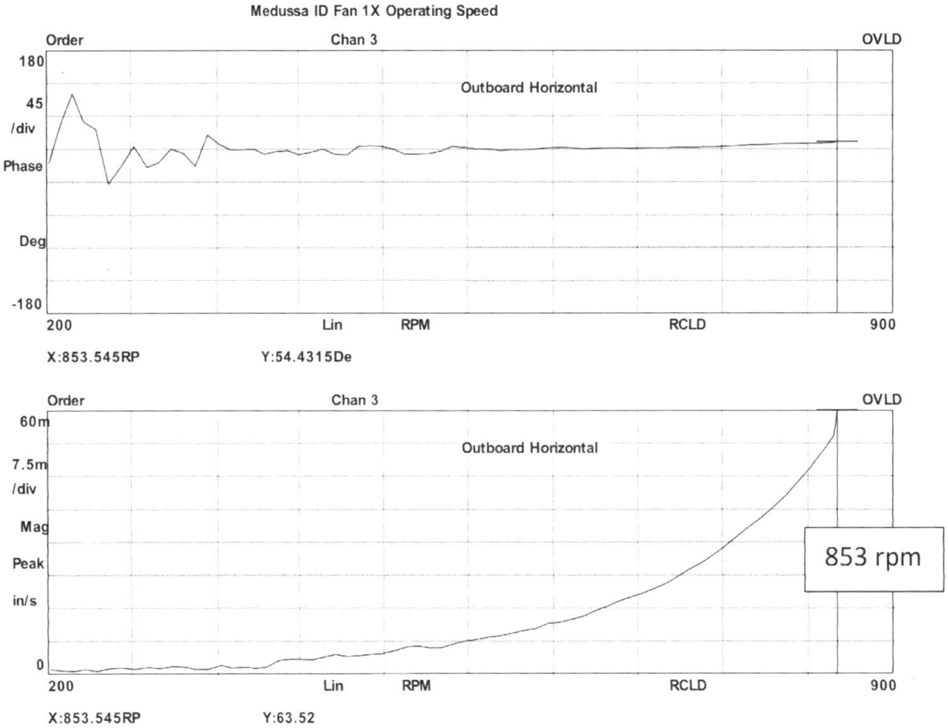

So, an associated technique is to review the 2nd order of vibration or twice the operating speed frequency as a function of rotor speed. Figure 13 below shows the 2nd order for the same machine and location shown in Figure 12. Figure 12 indicates a peak and phase shift at 557 rpm that the 2X response peaks, so therefore when the speed is at 557 rpm the response at twice the operating speed is amplified, therefore the natural frequency exists at 557 x 2 = 1104 rpm which is approximately 200 rpm above nominal operating speed.

Figure 13 : 2X Order Tracking Bode Plot (Vibration and Phase vs Speed) Nominal Operating Speed = 900 rpm

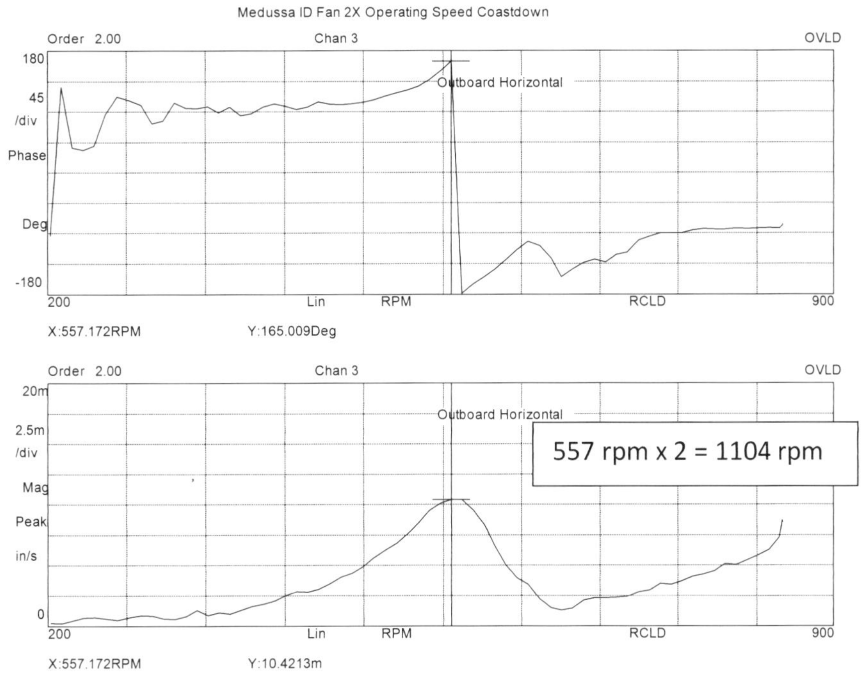

ID Fan Foundation Natural Frequency Identification Using Mechanical Shaker:

To identify the natural frequency of components or structures above the operating speed or machines that stop quickly after turning off, the mechanical shaker testing through the operating speed range and above is very beneficial. Typically, this results in a harmonic sweep of the shaker through the frequency (speed) range of interest. If the vibration response is identified to indicate amplification as in a natural frequency and mode, then the sweep can be stopped at that speed and a modal test can be completed similar to an operational survey technique using discrete measurement points in 3 dimensions or video methods.

This approach can be used for any number of speed ranges and sweeps in different shaker force directions. In this case, the shaker is positioned in a manner to excite the rocking modes of the fan foundation, similar to that deflection if the fan were being exciter by the fan wheel during operation.

As shown below, the 1X order tracking is measured up to 1150 rpm where the top fan design speed is operating at 900 rpm nominal. The plot shows a resonance condition at 1000 rpm when the fan is operating at 900 rpm, maybe too close for comfort.

Figure 14 : Natural Frequency Identified Above Operating Speed (Nominal Speed = 900 rpm)

Identified Natural Frequency ~ 1000 rpm

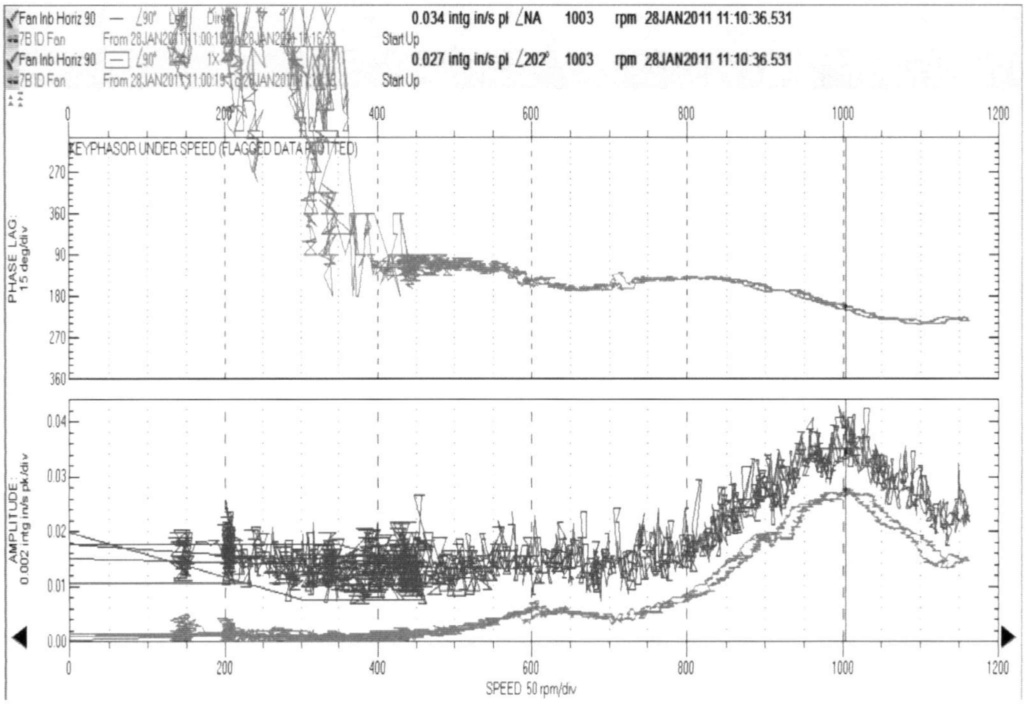

Another ID fan is tested similarly to that shown above, which also indicates a natural frequency close to the operating speed of 900 at 1130 rpm.

Figure 15 : Shaker Testing Foundation Response

(Plot Vibration Amplitude and Phase vs Shaker Harmonic Speed during Speed Sweep)

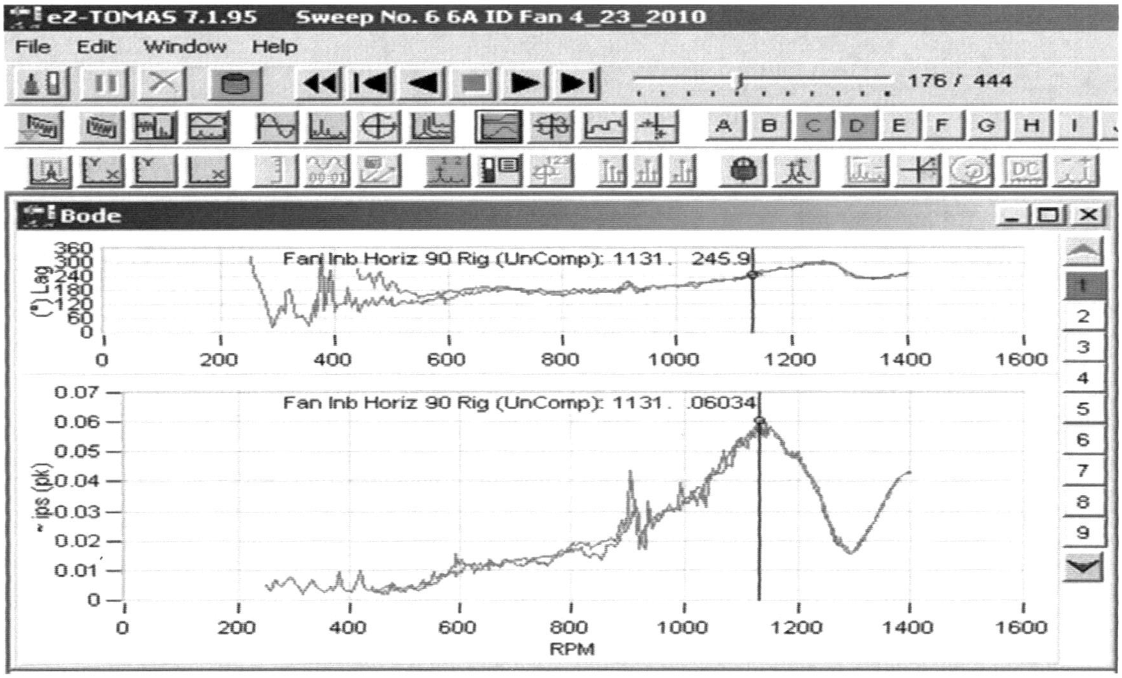

Figure 16 : Harmonic Shaker Attached to Foundation Horizontal Direction

Figure 17 : Fan Foundation with Motor Pier

Figure 18 : Harmonic Shaker Mounted to Top of ID Fan Foundation

Figure 19 : Fan Foundation Shaker and Expert

Figure 20 : Variable Speed Motor and Controller with Umbilical Drive

Figures 16 to 20 provide the photographs of the mechanical shaker that consists of two eccentric arms with weighted hammers. The eccentric arms counter-rotate in opposite directions to allow cancellation of forces in certain directions and to concentrate additive forces in other directions. The shaker is driven by a variable speed motor with a speed controller capable of driving the unit through a sweep of speeds up to 1800 rpm.

The mechanical shaker can be used to sweep through a speed range looking for amplified regions. Once the sweeps are completed, the mechanical shaker can be stopped at intermediate speeds while the vibration response is collected at various locations on a machine or structure. If the harmonic shaker is stopped on the natural frequencies, the mode shape can be extracted using a deflected shape software.

Impact Hammer Test on Large ID Fan Foundation:

Figures 16 and 17 below provide the photographs of attempts to use a large impact hammer to exciter the ID Fan foundation shown in Figure 16. A large pendulum mass suspended by steel cables was fixtured in a way to allow a "dead mass" mounted behind the impact test hammer force cell to measure the impact and create the impact hammer force. This method failed miserably for several reasons including;

1. The pendulum couldn't be retracted very far in a swing motion
2. The pendulum couldn't be controlled to impact and stay on target
3. The hammer-force cell was damaged due to the miss hits.

Figure 21 : Attempt at Forced Response Using Impact Hammer with Added Mass on Fan Foundation

(12 lb Hammer Head with Pendulum Mass Attached)

Figure 22 : Attempt at Forced Response Using Impact Hammer with Added Mass on Fan Foundation

(12 lb Hammer Head with Pendulum Mass Attached)

Impact Testing Examples:

Figures 23 to 25 provide plots and photographs of a gas turbine support frame that is used to support the gas turbine. The gas turbine has a vibration controls interface that was indicating spiking of the vibration at low frequencies. The interface uses displacement as the active units for machine protection and trip function if the vibration amplitudes exceed a certain displacement in mils p-p.

The controls interface was designed to provide a high-pass filter above 40 hz, with the operating speed of the gas turbine engine above 5000 rpm. It was determined from testing that the spiking was occurring at 8-9 hz.

When impact testing was completed, it was found that a natural frequency of the gas turbine trunnion support at 8.5 hz was found. This natural frequency is believed to be amplifying the response of the vibration probe due to aerodynamic flow noise thus creating the higher displacement.

The high-pass filter was later found to be switched off due to a glitch with a power transfer relay.

Figure 23 : Impact Testing of Gas Turbine Trunnion Support indicating 8.5 hz Natural Frequency

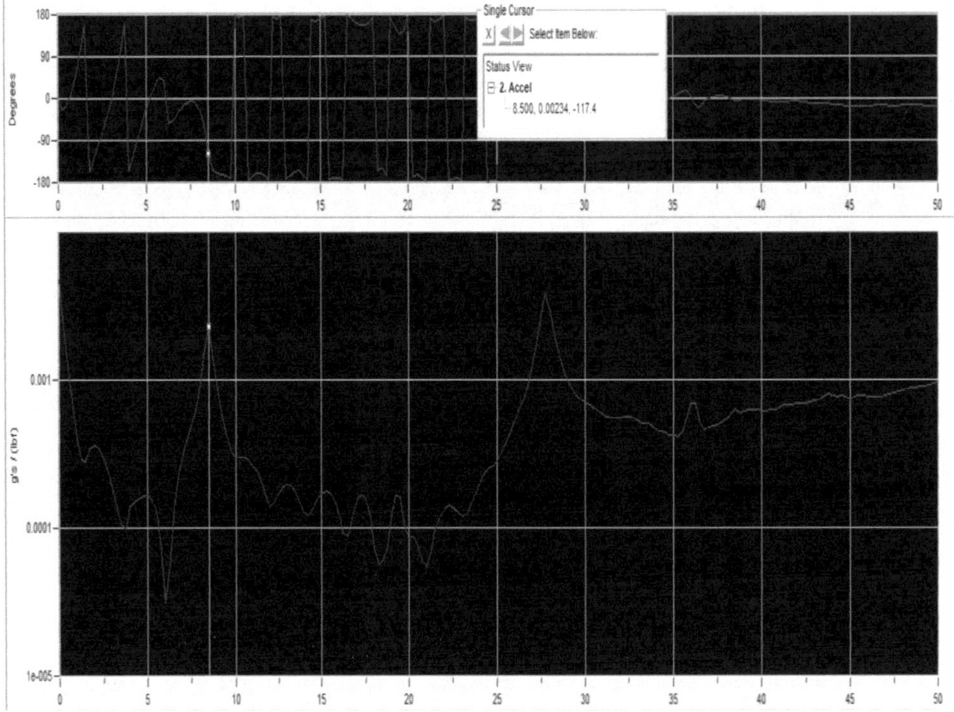

Figure 24 : Gas Turbine Trunnion Support indicating 8.5 hz Natural Frequency

Figure 25 : Gas Turbine Trunnion Support indicating 8.5 hz Natural Frequency

Figures 26 and 27 provide an example of a resonant condition where the boiler feed pump operating speed range was exciting a yawing mode between the two ends of the pump at the high-speed operation. It was determined that the yawing mode natural frequency could be reduced below the normal operating speed range. This "softening" of the pedestal was able to alter the stiffness to change the natural frequency of the yawing mode of the boiler feed pump.

Figure 26 : Boiler Feed Pump with Slots Cut in Support Pedestal to Detune Resonance at Upper Speed Range

Figure 27 : Boiler Feed Pump with Slots Cut in Support Pedestal to Detune Resonance at Upper Speed Range

Figures 29 to 31 provide an example of a resonant condition where the natural frequency was too close to the operating speed. The top of the motor was identified as a location that additional mass could be added to lower the natural frequency below the operating speed. Figure 29 provides the final configuration after some trial masses were added as shown in Figure 30.

Figure 28 : Vertical Pump Resonance with Fabricated Mass Top Hat

Figure 29 : Temporary Mass Addition during Testing of Vertical Pump

Figure 30 : Permanent Mass Addition during Testing of Vertical Pump

Bibliography

1. Rotor Dynamics and Balancing Course Notes – Syria, Virginia 1992
   (Jackson, Gunter, Eshleman)

2. 2003 Vibration Institute Summer Meeting
   Phase Angle Continuous Monitoring Detects Blade Failure (Lantrip, Woodson)

3. IMAC International Modal Conference 1992
   Diagnosing and Analysis of Resonance Problems on Rotating Machinery
   (Garrett, Lantrip)

4. July 2012 Vibration Institute National Training Conference
   Techniques used in Field Balancing (With and Without Rubs) (Lantrip)

5. Identification of Structural Characteristics of Prestressed Concrete Poles using Modal Testing Methods – Master's Thesis: University of Alabama in Birmingham Department of Civil Engineering (1995) Lantrip

UTILIZING ELECTRICAL SIGNATURE ANALYSIS TO DETERMINE ENERGY LOSSES IN ELECTRICAL AND MECHANICAL DEFECTS OF ELECTRIC MACHINES AND POWERTRAIN

Author: Howard W Penrose, Ph.D., CMRP

UTILIZING ELECTRICAL SIGNATURE ANALYSIS TO DETERMINE ENERGY LOSSES IN ELECTRICAL AND MECHANICAL DEFECTS OF ELECTRIC MACHINES AND POWERTRAIN

Howard W Penrose, Ph.D., CMRP
MotorDoc® LLC
Lombard, IL 60148
howard@motordoc.com

Abstract: One of the challenges when performing prognostics and diagnostics on equipment is the ability to both determine the severity and to prompt action to correct the defects. In electrical signature analysis the airgap torque or Kilowatts presented as linear spectra can be used to evaluate the individual defect losses. In addition to identifying the percentage losses versus total power to determine severity, the ability to perform a system-view of defects, the severity and direct energy cost impact of defects is important. In this paper we will discuss how the torque or kilowatt spectral data can be used to determine the energy losses across a defect to assist in presenting a directly measurable value that can be used to confirm corrective action success. The values can also be tied to greenhouse gas emissions to support any corporate carbon goals.

Keywords: Greenhouse Gas Emissions; Electric Machine; Energy; ESA; Kilowatts; Powertrain; Torque

Introduction

Electrical Signature Analysis (ESA) provides a method of evaluating a complete electric machine system from incoming power quality to the driven equipment (Figure 1) using the electric machine air gap as the transducer. ESA is a combination of Current Signature Analysis (CSA) and Voltage Signature Analysis (VSA) that provides the ability in an electric motor system to see both upstream (VSA) and downstream (CSA) from the point of test. The combined test uses the data from voltage and current to calculate power quality and has the ability to calculate power in watts and torque.

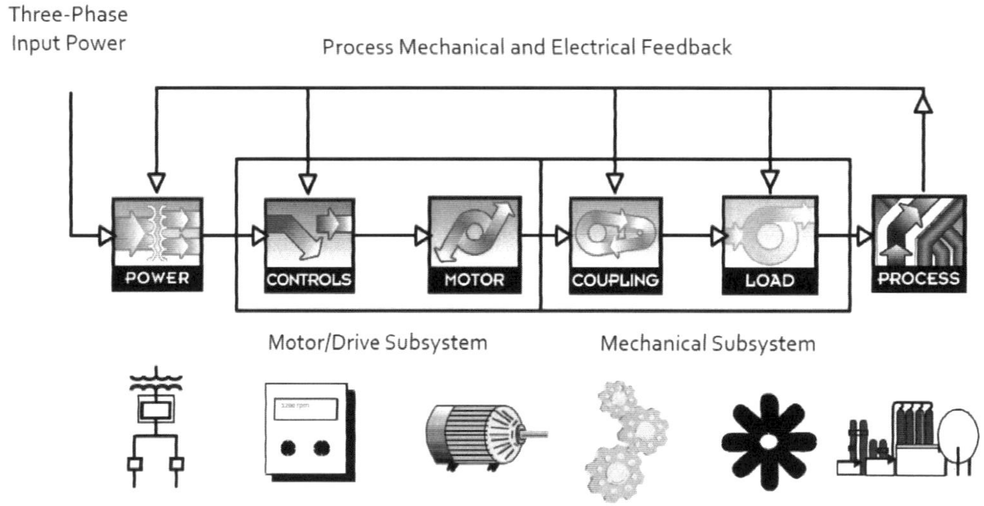

Figure 1: Electric motor system from incoming power to driven equipment and process

For generators VSA is used to evaluate the condition of the electric machine and drive train while CSA is used to evaluate downstream towards the load, with a few special exceptions such as Doubly Fed Induction Generators (DFIG) in wind power. In electric motors and generators, the airgap senses change due to torsional oscillation and radial movement of the rotor in the airgap. Due to the impact of small variations in the magnetic field on voltage and current values, variations of physical movement of the space between the rotor and stator are detectible in relative strength of the field (dB) and defects in the materials, such as rotor bars, stator slots, and laminations, also impact the fields. For mechanical and connected components, the 'drive-through' (motors) or 'ride-through' (generators) that cause variations in torque are also measured through the air gap. In an electric machine, each defect can be tied to an electrical loss and resultant minor drop in energy efficiency of the motor, for machine defects, and of the system, for power quality or powertrain defects.

For static and dynamic eccentricity, the variations in the airgap result in Unbalanced Magnetic Pull (UMP) which can be a measurable value when certain parameters of the electric machine are known. As identified by Kim et.al. [1], the closer the distance of one part of the rotor to stator of an electric machine, the greater the magnetic pull between the rotor and stator. These values can be used to determine the severity of certain conditions, such as misalignment, including soft foot conditions, or belt tension.[2] Additionally, rotor-related conditions are well documented from rotor bars to material defects [3][4] including wound-rotor and salient-pole machines.

The sensitivity of ESA to minor defects is also documented, such as winding stress prior to winding shorts [5] and missing wedges in both electric motors and generators, with wedge-type having an impact on the signatures, and coil movement including improperly supported coil ends.[6] Mechanical conditions such as bearings in the machine can be detected by component calculations for ball and roller bearings, and air-gap analysis with babbitt bearings.[7]

Misalignment is a primary challenge of ESA, as the primary result of unbalance or misalignment is a 1X running speed sideband of the power frequency. However, as the equipment extends beyond the coupling, the belt frequencies, sheave frequencies, and driven components can all be evaluated with ESA.[8] The values are consistent such that the amount of energy required to overcome defects beyond the coupling are measurable in the primary driver or driven electrical to mechanical converter.

The voltage and current values are converted to both a watt measurement (power) and torque measurement (Nm or lbft).[9] These may then be processed as spectra. One of the primary differences between the voltage and current spectra and the power and torque spectra are that the power and torque result in demodulated values while voltage and current are amplitude modulated. This means that VSA and CSA are produced as amplitude modulated values that are converted to dB values that are counted down from the peak of the respective voltage and current (Figure 2) while power and torque are measured from '0' up to their peak values (Figure 3).

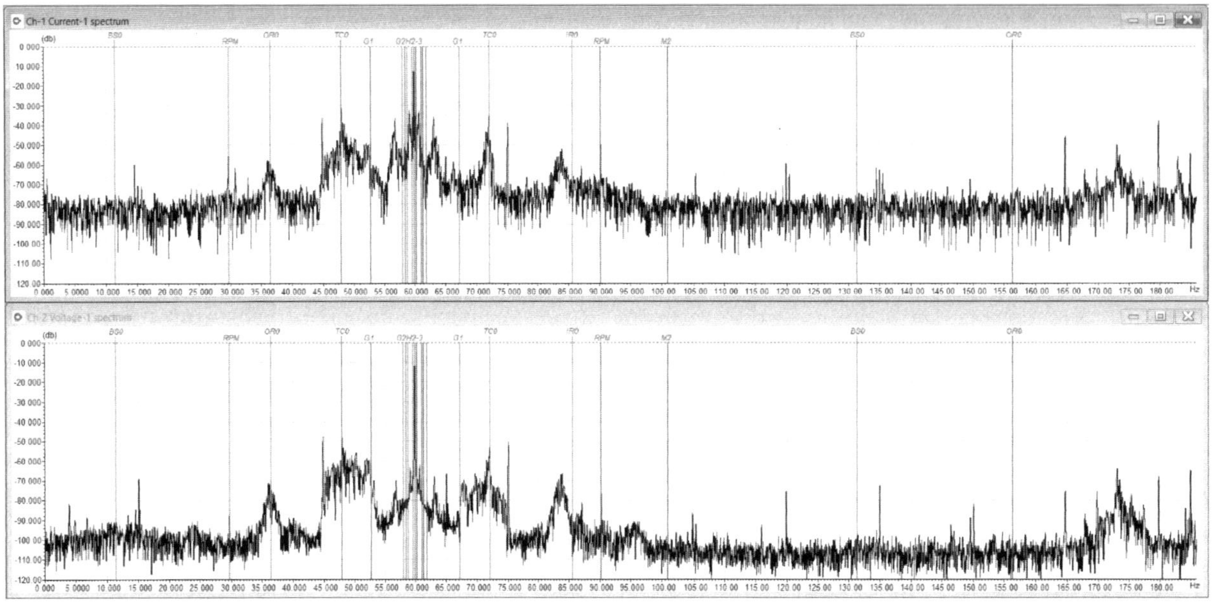

Figure 2: Current (top) and voltage (bottom) spectra showing peak line frequency at ~60Hz and mirrored sidebands

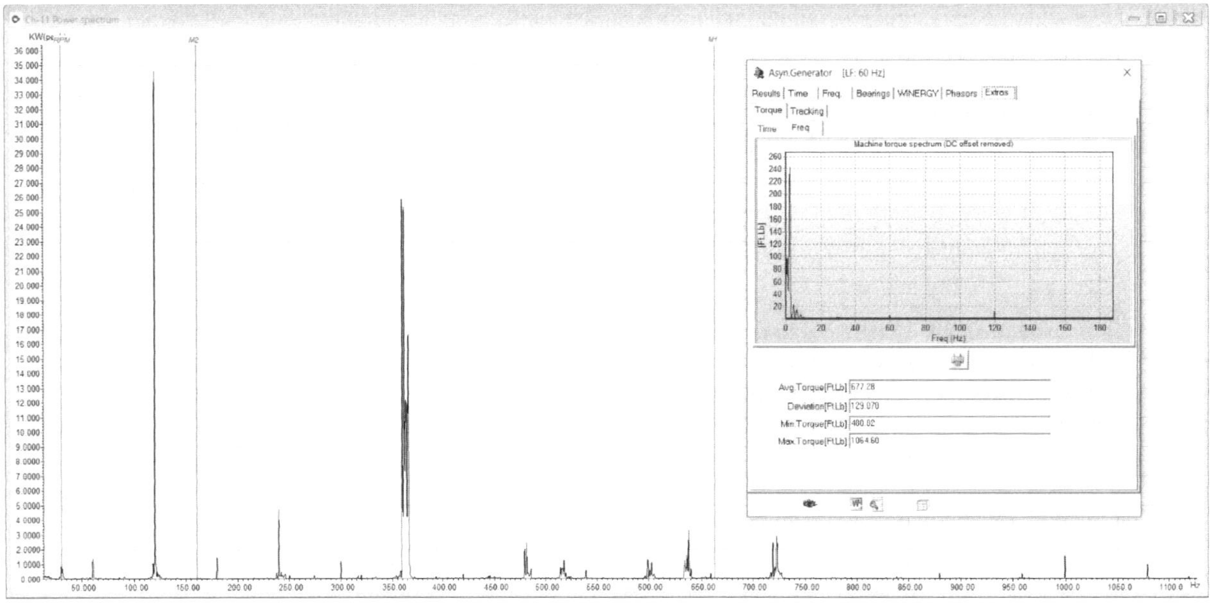

Figure 3: Power in Kilowatts and torque in ftlb (inset)

Detection and Calculation of Losses

Once a defect is identified using ESA then the loss value can be determined using either calculated torque or kW for each peak associated with the defect. Traditionally, the values that would trigger an evaluation are -45dB for induction motor rotor bar defects (the value varies by type of machine and application), -65dB (or 10dB greater than the noise floor) for induction motor other defects except bearings and stator defects, and any value over 5dB above the noise floor for bearings and stator defects. Certain special applications, such as wind turbine generators, have different values and related patterns.

To then be able to calculate the losses in real dollars you will need to know the hours per year of operation, the demand cost, and the usage cost for the equipment. This value can vary by season. Next, you will need to determine any variation in loading for the motor or equipment, if you want to show before and after costs for operating the machine (not recommended). Finally, for the Greenhouse Gas Emission (GhG) value, you can either use the 2019 generic USA 0.707 metric Tons CO_2/MWh or use the EIA.gov or EPA.gov values for the associated region, which considers power generation type.

For example: a 181-kW load operating over 6000 hours per year, with a \$10/kW demand and \$0.10/kWh usage cost would be calculated as shown in eq. 1, 2, and 3.

181kW x \$10/kW x 12 Mo. = \$21,720 demand (eq. 1)

181kW x \$0.10/kWh x 6000 hr = \$108,600 usage (eq. 2)

Total Cost (energy) = ~\$130,320/yr (eq. 3)

If we then want to demonstrate the GhG emission related to the application, we would calculate it as shown in eq. 4. For this example, we will use the 0.707 Tons CO_2 per MWh.

$$\left(\frac{181kW \ x\ 6000hr}{1000kW/_{MW}}\right) x\ 0.707\ ^{tons\ CO_2}/_{MWh} = \sim 768\ Tons/Yr \qquad \text{(eq. 4)}$$

For smaller machines, or total kWh losses, the value may be presented in pounds or kilograms of CO_2. For this paper we will use the 2019 EIA.gov value of 1.56 lbs of CO_2 per kWh. In this case, we can use the example of a defect found to be 5 ftlb at 16Hz in a machine operating 4000 hours per year and apply it to a simple torque formula (eq. 5) which results in ~4,255 lbs CO_2 per year.

$$lbs_{CO_2} = \frac{(16Hz\ x\ 60^{RPM}/_{Hz})x5ftlb}{5252} x0.746\ ^{kW}/_{HP}\ x1.56\ ^{lbs_{CO_2}}/_{kWh}\ x4000\ hrs \qquad \text{(eq. 5)}$$

The appropriate selection of the above values will depend on how your company or customer reports out carbon emissions. It is recommended that you convert to tons CO_2 when you exceed 10,000 lbs worth of losses. The loading is also critical as the values will change as the loading changes. For variable loads, calculate for several load points and hours, or for the average loading based upon when you have tested.

Case 1 Belted Application

A 300 hp motor is detected with belt wear from belt aging and turbulence related to incorrect settings in the ventilation system (Figure 4). The torque values are reviewed (Figure 5) and the belt peak is found as 5ftlb at 11 Hz and the turbulence is 48ftlb at 1.5 Hz with \$10/kW demand, \$0.10/kWh, and 6000 hours of constant load operation. By plugging the numbers into (eq. 1 – 3) the result is ~\$338 for the belt per year and ~\$439 for the turbulence for a total of ~\$777 in energy costs and ~10,130 lbs CO_2 emissions (~4.6 Tons). With this load being 181 kW (see above example), this represents a 0.6% improvement in efficiency and 0.6% improvement in GhG.

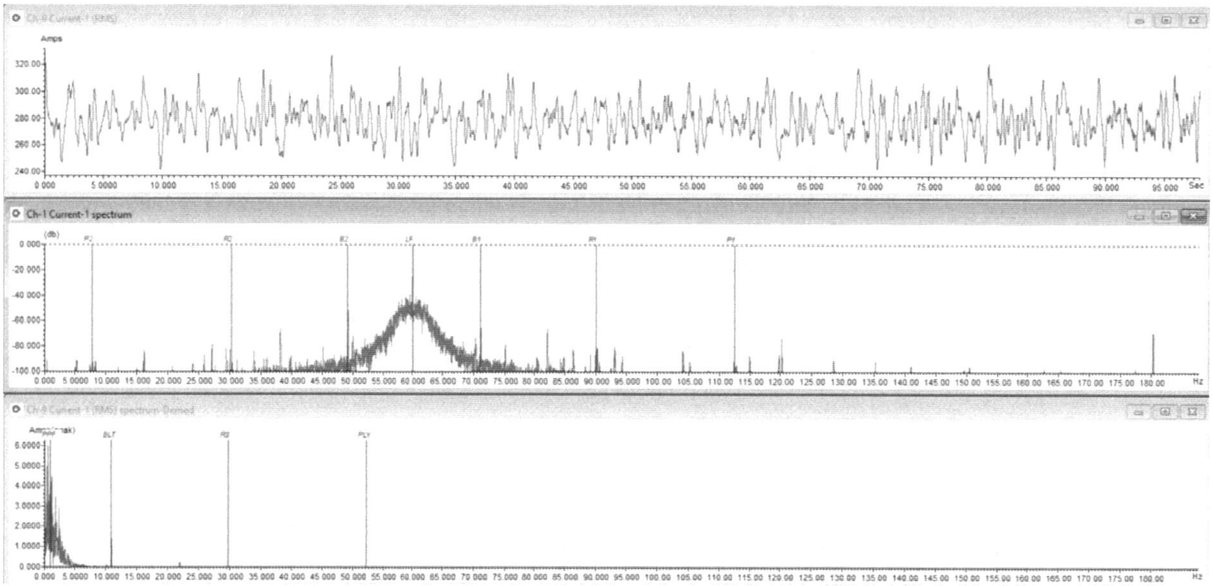

Figure 4: belt wear and turbulence on a 300 hp motor operating at 181 kW

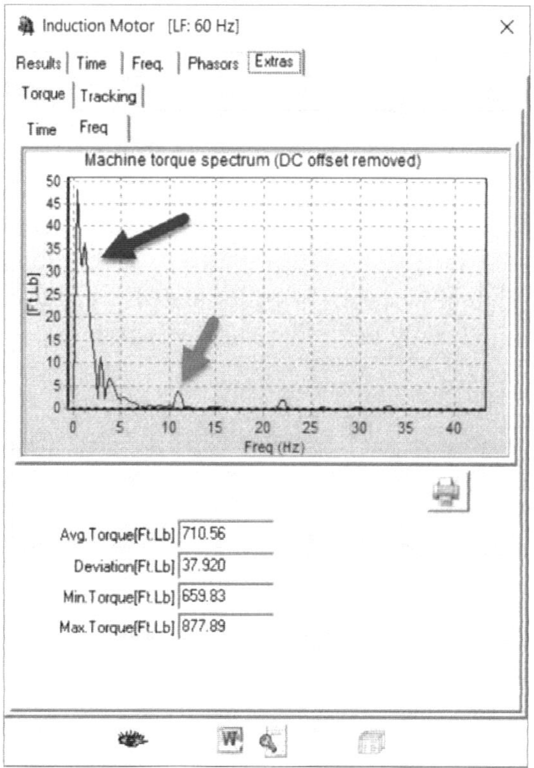

Figure 5: Purple is the turbulence and red is the belt frequency.

Case 2 150 Horsepower Motor with Belt Misalignment and Mistensioning

A 150-horsepower motor operating at 92% load and 100 kW is found to have loose belts and significant misalignment (Figure 6). The peak values are found to be 60ftlb at 9 Hz for belt slip and 23 ftlb at 16 Hz for the misalignment (Figure 7). Using the same energy values from Case 1, the result is a total energy cost of $3,312/yr (4.6kW) for the belt slip and $2,232/yr (3.1kW) for the misalignment. Additionally, at a total of 46,200 kWh, the GhG emissions can be presented as either ~72,072 lbs CO_2/yr or ~33 Tons CO_2/yr. In this case, the drop of 7.7 kW load when corrected improves the system efficiency from 100 kW to 92.3 kW for a 7.7% improvement.

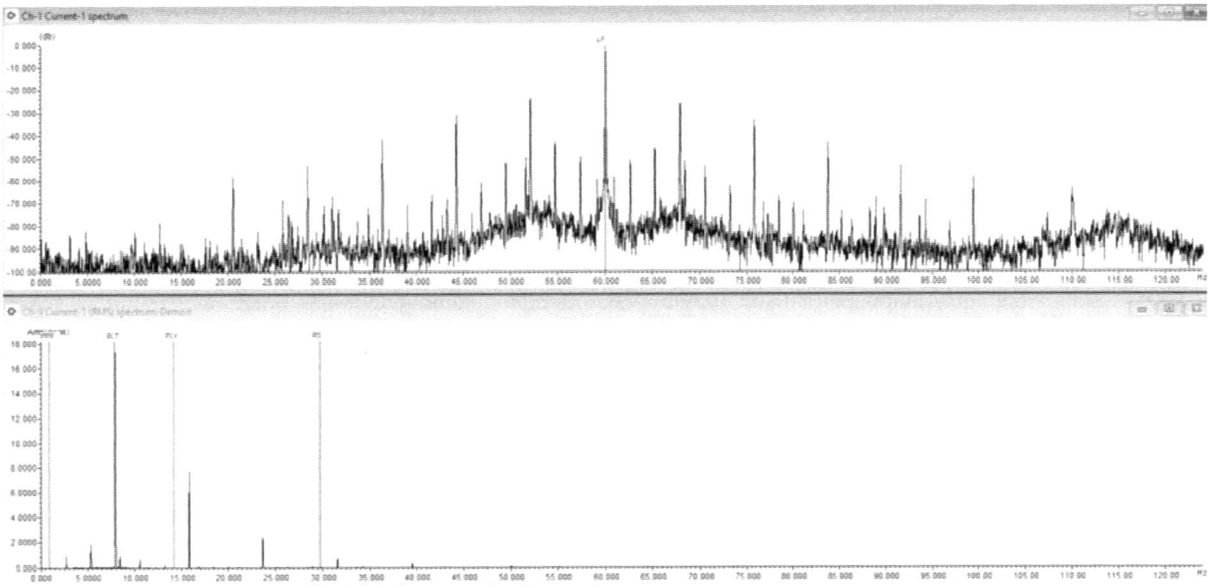

Figure 6: Belt, misalignment, and other frequencies on a 150 hp motor

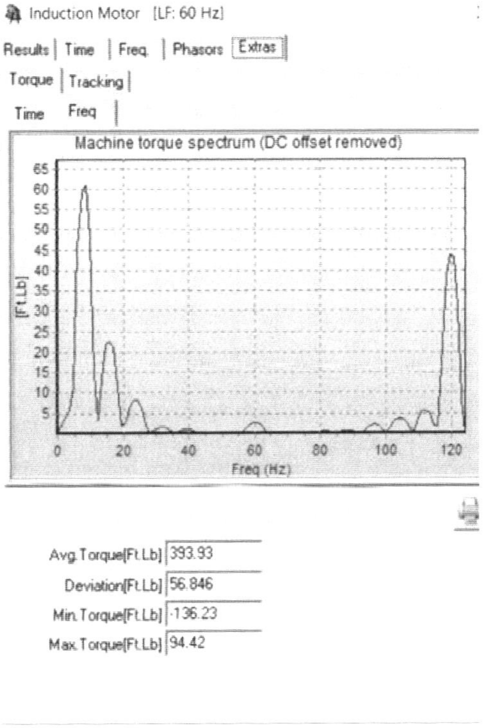

Figure 7: Torque peaks associated with Case 2 and Figure 6

Case 3 Pump Motor with Bearing and Impellor Defects

A 600 hp pump and motor (Figure 8) on a VFD was found to have a defective bearing and impellor during a commissioning test. The bearings were SKF 6322 on both the drive end and opposite drive end with 6 blades on the impellor. The motor load at time of test was 27%, or 114.5kW with other conditions also found in the machine which was operating at 42 Hz line frequency from the drive (Figure 9). The kW peaks were found as shown in Figures 10 and 11, which are totaled as:

- Bearing Cage: 0.25kW
- Bearing Ball: 0.58kW
- Bearing Outer Race: 0.44kW
- Pump Blade: 1.32kW and 2.9kW
- Total = 5.49 kW

Figure 8: 600 hp VFD motor and pump application

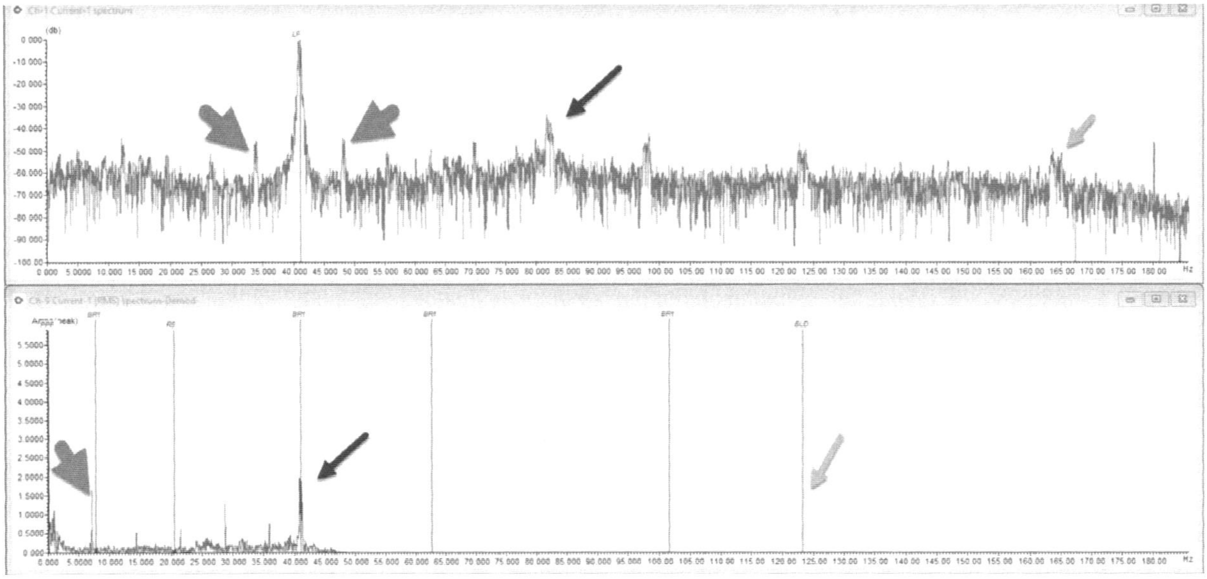

Figure 9: Fault detection on 600 hp motor

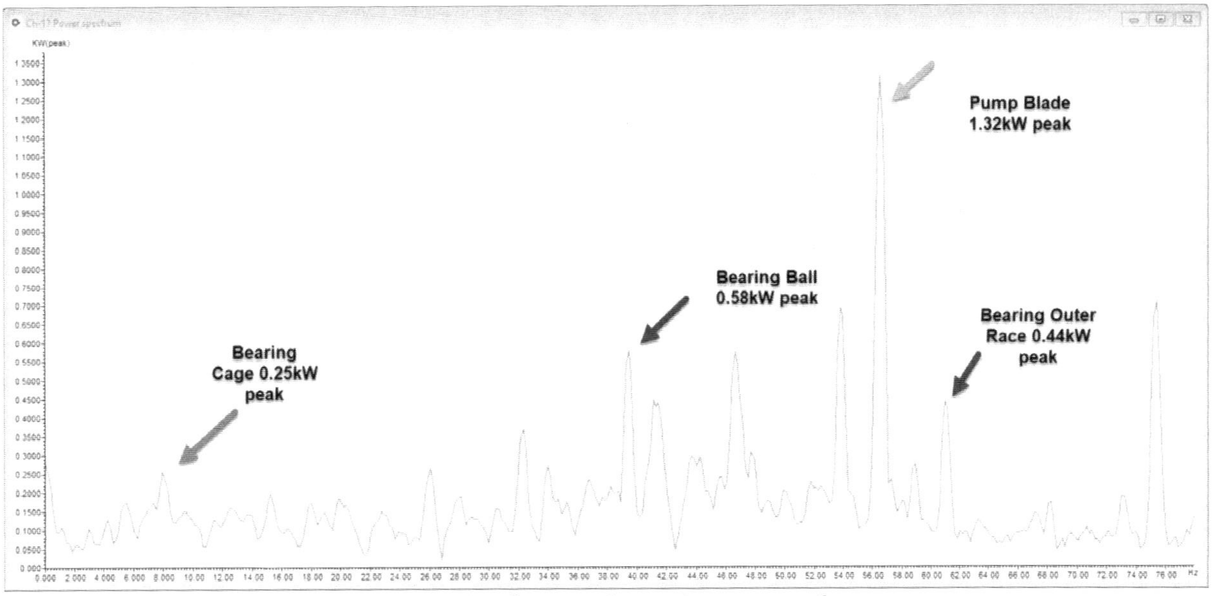

Figure 10: Bearing frequencies and one pump frequency

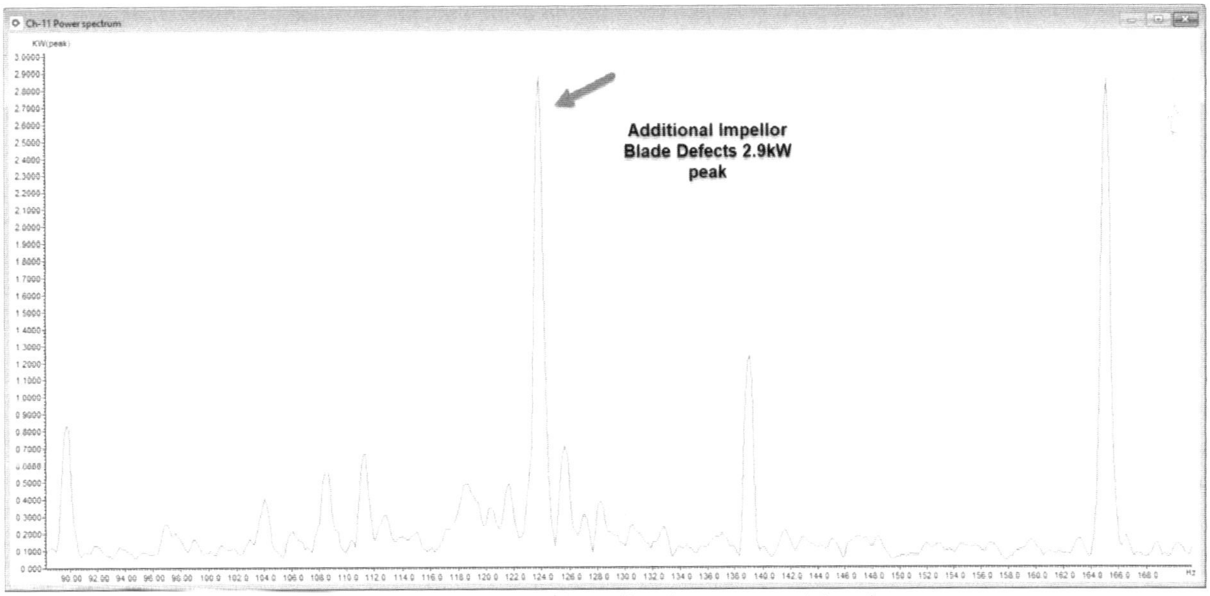

Figure 11: Additional impellor frequency with value

This application operates 8760 hours, although has a variable application in which the point of test was during the lightest load condition. The total when applied to the above equations at the lightest load is ~\$5,469/year and 38.5 tons CO_2. Once the application was corrected, the peaks no longer existed, and the motor and pump ran quieter. Overall, the impellor and bearing defect resulted in a 4.8% loss in system efficiency, which was corrected.

Conclusion

The application of ESA as a system analyzer is beneficial. The additional ability of ESA over CSA/MCSA and VSA only includes the ability to view the complete system and calculated associated losses with energy costs and greenhouse gas emission values. The use of the additional tools can assist in supporting corrective actions on findings including the ability to verify the impact of the correction as part of commissioning the equipment back into service.

Bibliography

[1] H. Kim, J. Nerg, T. Choudhury and J. T. Sopanen, "Rotordynamic Simulation Method of Induction Motors Including the Effects of Unbalanced Magnetic Pull," in IEEE Access, vol. 8, pp. 21631-21643, 2020, doi: 10.1109/ACCESS.2020.2968915.

[2] Penrose, H. "Electrical Signature Analysis Air-Gap Eccentricities (Part III): Belted Applications," TheRamReview.com, Nov. 29, 2020, https://theramreview.com/electrical-signature-analysis-air-gap-eccentricities-part-iii-belted-applications/

[3] William T. Thomson, Ian Culbert, "Current Signature Analysis for Condition Monitoring of Cage Induction Motors: Industrial Application and Case Histories," Wiley-IEEE Press, 2017.

[4] Penrose, H. "Regarding Reliability of Electric Motors: Trending Rotor Bar Failures," TheRamReview.com, Sep. 4, 2020, https://theramreview.com/regarding-reliability-of-electric-motors-trending-rotor-bar-failures-case-study/

[5] K. Alewine and H. Penrose, "Field Experiences Utilizing Electrical Signature Analysis to Detect Winding and Mechanical Problems in Wind Turbine Generators," 2020 IEEE Electrical Insulation Conference (EIC), 2020, pp. 246-248, doi: 10.1109/EIC47619.2020.9158573.

[6] H. W. Penrose, "Evaluation of Stator and Rotor Interturn Stress with Electrical Signature Analysis in Variable Frequency Drive and Wind Generator Applications," 2020 IEEE Electrical Insulation Conference (EIC), 2020, pp. 274-277, doi: 10.1109/EIC47619.2020.9158738.

[7] Penrose, H. "Electrical Signature Analysis and a Wind Turbine Generator Bearing," TheRamReview.com, Jan 9, 2021, https://theramreview.com/electrical-signature-analysis-and-a wind-turbine-generator-bearing/

[8] Penrose, H. "Electrical Motor Diagnostics: 2nd Edition," Success by Design, 2008, Old Saybrook, CT.

[9] B. Lu, T. G. Habetler and R. G. Harley, "A Nonintrusive and In-Service Motor Efficiency Estimation Method using Air-Gap Torque with Considerations of Condition Monitoring," Conference Record of the 2006 IEEE Industry Applications Conference Forty-First IAS Annual Meeting, 2006, pp. 1533-1540, doi: 10.1109/IAS.2006.256733.

Bio: Howard W Penrose, Ph.D., CMRP, is the president of MotorDoc® LLC, a Veteran Owned Small Business, a past-chair of SMRP, past-chair Chicago Section IEEE, and serves on the IEEE PES Materials Subcommittee standards group, and American Wind Energy Association technical standards group. He has over 35 years in the electrical machinery industry from academia to military, and repair to design. Howard is one of the world's leading experts in electrical and current signature analysis.

USING FINITE ELEMENT MODELS AS A DIGITAL TWIN FOR THE DEVELOPMENT OF VIBRATION CORRECTIVE ACTION PLANS

Author: Robert J. Sayer, PE

USING FINITE ELEMENT MODELS AS A DIGITAL TWIN FOR THE DEVELOPMENT OF VIBRATION CORRECTIVE ACTION PLANS

Robert J. Sayer, PE
Applied Structural Dynamics
Westerville, Ohio
sayerinc@bright.net

Abstract: A ***digital twin*** is a virtual representation of a living or non-living physical entity. It can serve as a real time digital counterpart of a physical object or process. Using a digital twin, not only can different failure scenarios be studied, but corrective action plans can be evaluated prior to implementation. This presentation will concentrate on the finite element analysis (FEA) method, which is one of the earliest forms of representing a physical entity using a numerical model.

The finite element analysis is an approximate numerical technique. Early finite element codes were developed by NASA to support the space program, providing stress, dynamic response, and heat transfer analysis of complex systems. This paper presents the evolution of using finite element simulations for root-cause failure analysis of industrial mechanical equipment, and to assist in the development of corrective action plans. Special emphasis is placed on correlation of the digital twin to test data. Case studies spanning nearly four decades will be used to illustrate the progression of the technique.

These cases involve:

1) Fatigue failure of large autogenous grinding mills at an iron ore mine and processing plant performed in the early days of finite element analysis,
2) Extremely reduced mean-time-between-failure (MTBF) rates of belts for cooling tower fans on an offshore production platform performed recently using modern FEA capabilities, and
3) Catastrophic failure of a fan rotor at a food processing plant that required FEA to estimate the effects of centrifugal stress stiffening.

Key Words: Autogenous grinding mills, centrifugal fans, centrifugal stress stiffening, digital twin, fatigue, finite element analysis, notch-strain life, rotor dynamics

History of FEA

Finite element analysis is an approximate numerical method that enables the solution for the stress and dynamic analysis of complex structural-mechanical systems. The development of the FEA method was first attributed to Richard Courant in 1943. He proposed breaking a continuum into discrete triangular segments to perform stress analysis using the Raliegh Ritz method.

In the 1950's Boeing used triangular segments to perform stress analysis of airplane wings. Dr. Ray Clough introduced the term "finite element" to this numerical technique. The main drawback of the method was the computational requirements needed to solve even simple problems in continuum mechanics. However, many of these computational constraints were relaxed with the advent of the digital computer.

In the early 1960's, MacNeal Swendler Corporation (MSC) was contracted by NASA to develop a general-purpose finite element code, which was named NASTRAN. Shortly thereafter, several other codes, such as ANSYS and MARC, became commercially available. Computational requirements continued to be the major constraint to using the method. Most analyses were done using main frame computers on a timeshare basis which was expensive. The input consisted of a deck of punch cards. The ability to check the validity of a finite element model, prior to submittal to the main frame computer, was limited. The results of the analysis were typically provided as a ream of computer paper.

Today, static stress, modal, and forced dynamic response analyses of large finite element models can be done on a personal computer. Graphical interfaces are now available that assist the analyst in developing and checking a model and interpreting the results. This paper provides a case study of a FEA performed in the early period of finite element analysis and another case study that was done recently.

CASE STUDY #1:

Figure 1: Typical Autogenous Grinding Mill

A taconite plant had nine large autogenous grinding mills that reduced ore supplied from a primary crusher in order to concentrate the taconite and remove unwanted product. A typical autogenous mill is shown in Figure 1. The mills of this case study, consisted of a center cylindrical section (diameter = 36 ft) and conical sections at each end that transition to a smaller cylindrical section (diameter = 10 ft) that served as part of a hydrostatic bearing. During 1976, six mills were installed during the initial construction of the plant. The other three mills were installed in 1979 as part of a plant expansion. The supplier and design of the original mills were different than the mills installed during the expansion.

In early 1981, after less than two years of operation, the weldment between the main cylinder and the conical heads of all three of the expansion mills developed fatigue cracking that propagated completely through the shell causing product to leak from the inside of the mill. It was decided to develop a finite element model to serve as a digital twin to assist in the study of the probable root-cause of the failures. Finite element models of both mill designs were developed to compare the stress levels of the original mills to the expansion mills. The cylindrical-to-conical shell designs of each are shown in Figure 2. The original design contained a thicker transition piece. The conical end piece of the expansion mills were thicker than the original mills. However, the original mills had several stiffener plates (not shown) extending radially from the bearing trunnion to the outer diameter of the mill.

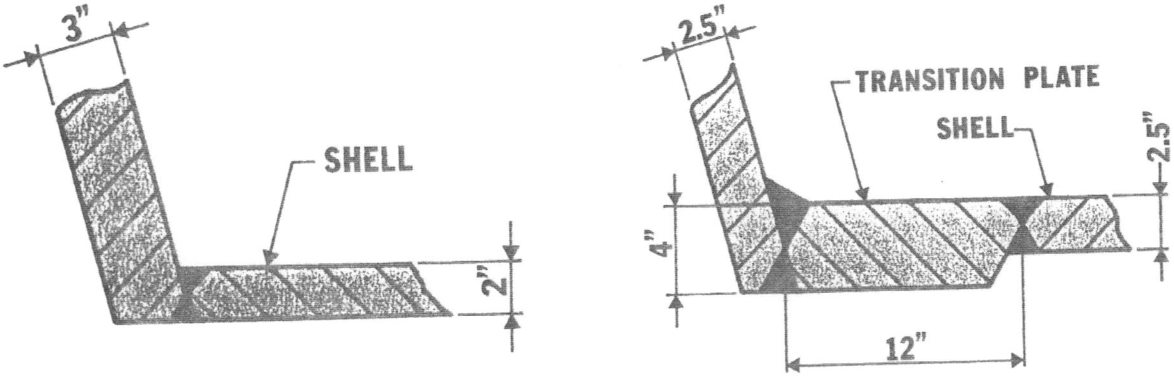

Figure 2: Expansion Corner Design (Left) versus Original Corner Design (Right)

During the time period that this study was performed, in order to conserve on computational requirements and minimize the cost of performing the analysis using a computer time-sharing service, the analysis was done using a two-step process. First, a global analysis of the entire mill was done using coarsely spaced elements that limited the total degrees-of-freedom of the model. This global model was coded using a hand sketch (Figure 3) for guidance. A more refined local FEA model (Figure 4) was then developed to estimate the concentrated stress level at the subject cylinder-cone intersection. Boundary conditions for this local model were obtained from the global model of the entire mill.

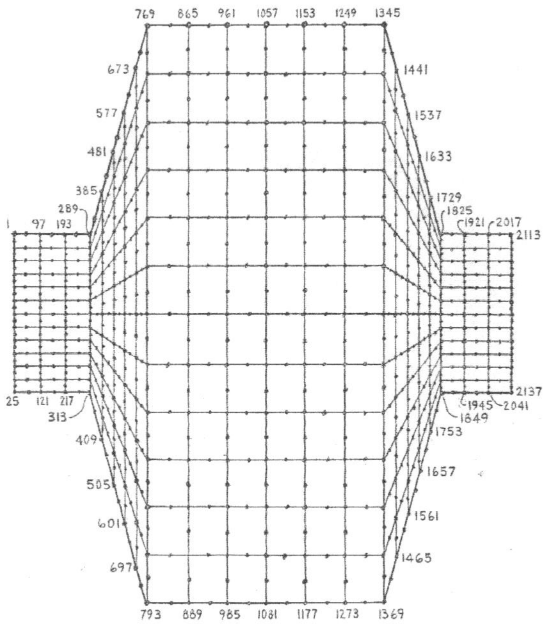

Figure 3: Sketch of Global FEA

An experimental stress analysis was performed on one of the expansion mills after a temporary weld repair was made to the fatigue cracks. Strain gages were installed to the cylindrical shell underneath the protective liners. Data was transmitted from the gages to an acquisition center using FM telemetry.

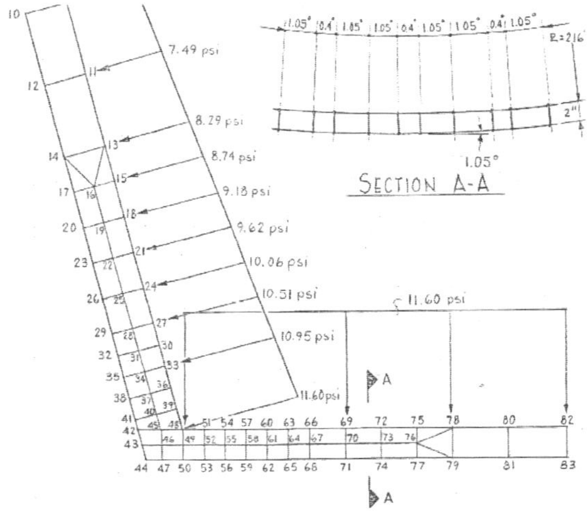

Figure 4: Sketch of Local FEA

The strain gage test confirmed that the stress cycle was dominated by the shell moving into-and-out-from the load zone of the charge. There was only a small amount of stress caused from the dynamic response of the shell to impacting within the charge or from dynamic forces transmitted to the shell from the large drive train. This information was important for the confirmation of the loading condition used in the finite element model.

Figure 5 shows a plot of the stress distribution predicted by the finite element model versus the stress distribution obtained by the strain gages. The stress range predicted by the numerical model (23,000 psi) was very close to the stress range measured by the strain gages (24,000 psi). The correlation of numerical prediction with actual measured results was very good.

This indicated that the FEA analysis procedure, loading conditions, and boundary conditions used for the numerical analysis provided an accurate approximation of the stress distribution through the subject weldment. With a minor adjustment to account for the small difference between calculated and measured stress, a digital twin was now available to study proposed design modifications. Also, an analysis procedure was confirmed that could be used to evaluate the shell design of the original mills to serve as a digital twin thereof.

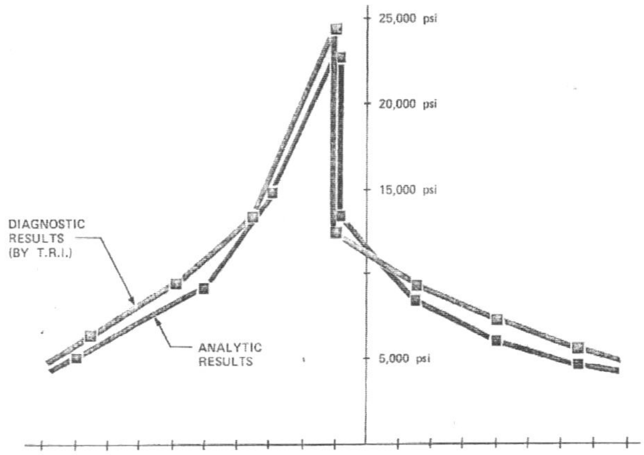

Figure 5: Calculated versus Measured Stress

Fatigue life curves, developed from constant amplitude fatigue life tests, can be described by a low-cycle fatigue component and a high-cycle fatigue component. Based upon this method, which is known as the *notch strain-life method*, the cyclic stress ($\Delta\sigma/2$) that will provide a desired number of fatigue life cycles (2N) can be estimated using Equation (1)

$$(\Delta\sigma/2) = (\sigma_f` - \sigma_m)(2N)^b + (\varepsilon_f`)E(2N)^c \qquad \textbf{Equation (1)}$$

Where $\sigma_f`$ is the fatigue strength coefficient, $\sigma_m$ is the mean stress, b is the fatigue strength exponent, $\varepsilon_f`$ is the fatigue ductility coefficient, and c is the fatigue ductility exponent. These are material properties. References [1][2][3] are available that provide values for these coefficients and exponents for various materials.

For cases of high-cycle fatigue, the fatigue life equation is dominated by the first term of Equation 1 and can be simplified to Equation 2.

$$(\Delta\sigma/2) = (\sigma_f' - \sigma_m)(2N)^b$$ **Equation (2)**

The expansion mill had undergone approximately eleven million cycles of stress when the fatigue cracking was discovered. At this same time, the original mills had experienced more than 28 million cycles without failure. The stress analysis indicated that the stress range of 24,000 psi consisted of a maximum tensile stress of 19,000 psi and a maximum compressive stress of 5,000 psi. The mean stress was 7,000 psi tensile.

The mill was made of carbon steel with properties that when input into Equation (2) provided the following equation for fatigue life:

$$(2N)^{-0.12} = (12{,}000)/(95{,}000 - 7{,}000)$$

This fatigue model predicted a fatigue life of around 12,000 million cycles, which in terms of life prediction, is considered a good estimate (predicted life = 3 years; actual life = 2 years).

The finite element model was then revised to serve as a digital twin for the original mill design in order to estimate the stresses that would be developed in the shell of the original mills. The results of this analysis are included in Figure 6 compared to the results for the expansion mill. An estimated maximum stress range for the original design was predicted at around 12,000 psi (tension = 9,000 psi , compression = 3,000 psi).

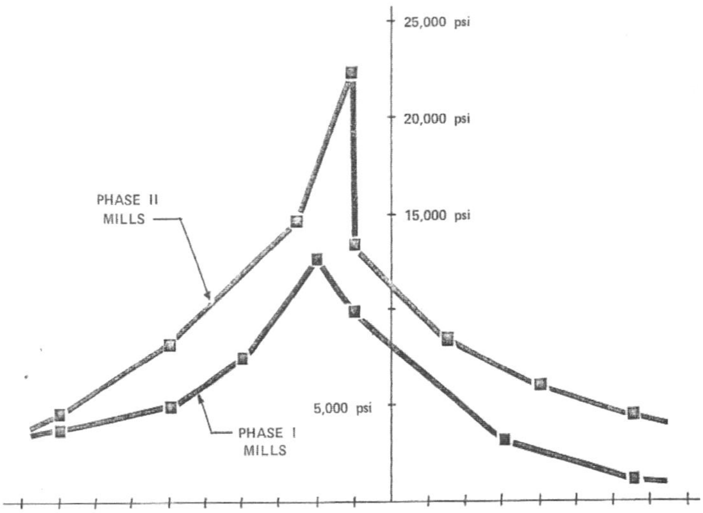

Figure 6: Stress Distribution (Original versus Expansion)

Using the same notch-strain fatigue life equation predicted that initial failure of the original mill could be expected in over 1 billion cycles or 166 years, well beyond the expected design life of the mill. Therefore, a design similar to that of the original mills was developed for the expansion mills, using the FEA model as a digital twin. It is interesting to note that these mills did operate for over 30 years without any fatigue issue.

CASE STUDY #2:

Figure 7: Photo of the FPSO

This case study involves the cooling towers on a deep-water FPSO (floating production storage and offloading) platform (Figure 7) that had experienced many belt failures and several shaft failures. FPSOs are used by the offshore oil and gas industry for processing hydrocarbons and for storage of oil. These cooling towers serviced three field gas compressors. Each compressor had three stages and each stage had two cooling towers, for a total of eighteen cooling towers. The cooling tower fans were belt-driven by 1450 rpm motors. Some of the fans were fixed-speed and others were variable-speed. The maximum operating speeds ranged from 242 to 272 rpm (4.03 - 4.53 Hz).

Figure 8 contains a sketch of the fan. It was vertically mounted and consisted of a shaft, impeller and drive sheave. The impeller was located above the upper bearing (cantilevered design). The upper bearing was a single-row, medium-duty, ball bearing. The lower bearing was a double-row, tapered roller bearing. These fans were forced-draft type, located at the bottom of the cooling tower, forcing air to flow upward through the cooling tower from the bottom. The bearings were supported from a flexible structure that spanned from side-to-side of the cooling tower.

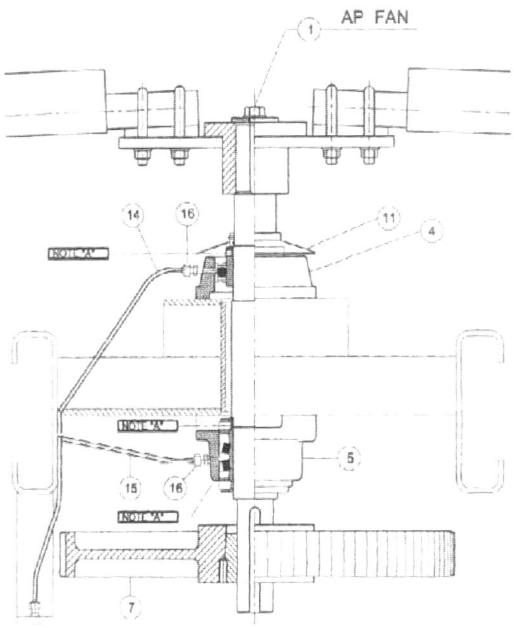

Figure 8: Sketch of Cooling Tower Fan

Figure 9 is a frequency spectrum of vibration data acquired from the fan bearing closest to the impeller. The spectrum was dominated by responses at the rotational speed of the fan and many of its harmonics. Notice that the vibration (0.006 ips) at 1x fan speed and the overall vibration (0.105 ips) were very low. These vibration levels did not suggest that a problem related to vibration existed. However, the spectrum, with an abundance of harmonics, indicated the possibility of a non-linear response, probably due to impacting.

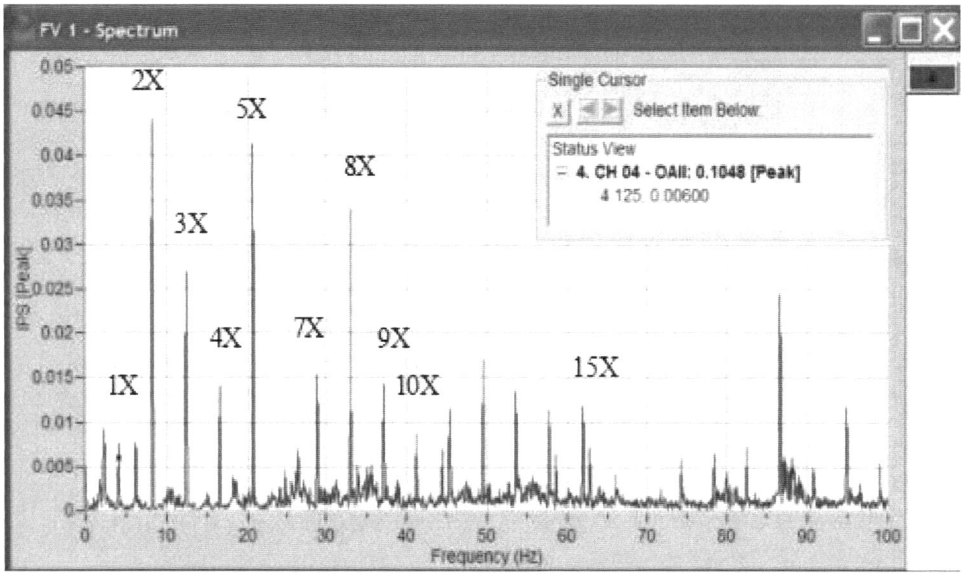

Figure 9: Frequency Spectrum of Bearing Vibration

Although the data indicated the presence of a non-linearity, it did not provide sufficient information to make a definitive diagnosis of the root cause of the belt failures. On the contrary, since vibration levels were low, the data could have led to an erroneous conclusion that the belt failures were not associated with vibration of the fan. This could have resulted in futile efforts to precision align the pulleys and/or adjust belt tensioning.

Figure 10 shows the results of an impact test performed to excite the natural frequency of the fan rotor. A natural frequency was clearly evident at around 4.5 Hz that was very close to the operating speed of the fan (4.125 Hz). The characteristic of the transfer function at this natural frequency suggested that the mode associated with it had a larger amount of damping than a typical structural mode. The peak was wide and rounded. The responses at the apparent natural frequencies at 18 Hz and 20 Hz were more representative of typical low damped structural modes. These peaks were sharp and narrow.

The impact test was performed in a manner directed at exciting the principal natural frequency of the fan rotor. Typically, the natural frequency of an overhung fan rotor in roller element bearings is expected to be a flexural mode of the shaft; the shaft bends between the bearings and the impeller wobbles as a result of shaft flexure. If the bearings provide substantial support and the shaft flexes, the nature of the response of the natural frequency to an impact test would be expected to resemble that at 18 Hz or 20 Hz, not that at 4.5 Hz.

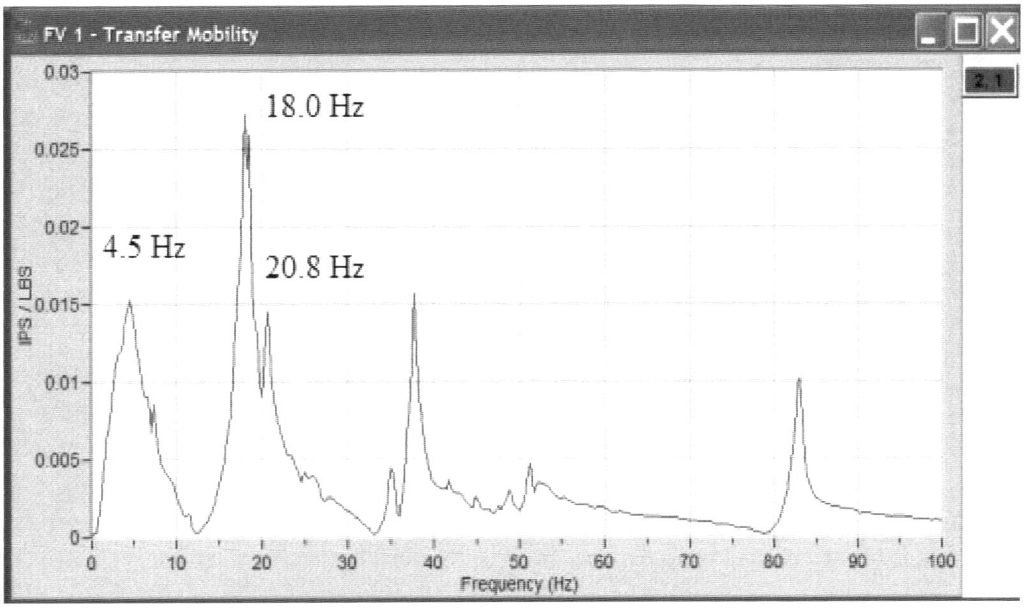

Figure 10: Natural Frequency Impact Test Results

The derating effect of bearing clearance on the stiffness of the rotor can have a marked effect on the modal characteristics of the rotor. There have been cases, several which have been documented by Sayer [4][5], where clearance in anti-friction bearings have resulted in the reduction in the natural frequency associated with the principal flexural mode of centrifugal and axial-vane fan rotors. The horizontal clearance in the bearings reduces the effective stiffness of the support, resulting in a reduction in the natural frequency.

The derating effect of bearing stiffness and clearances on the natural frequency of a rotor in anti-friction bearings is recognized in the "Handbook of Rotor Dynamics" [6] in which it is stated:

> "...to avoid compression of rolling elements and premature fatigue of rolling elements and/or races associated with differential thermal expansion ... bearings must be designed with radial clearance between the rolling elements and the races. Such bearing clearance may ... result in an effective softening of the bearing support stiffness and a very significant reduction in natural frequencies and critical speed. Moreover, the nonlinear nature of the effect of clearance on bearing support stiffness may cause many anomalies in vibration response to unbalance, such as hysteresis and jumps in vibration amplitude..., pseudo-critical speeds at whole-number multiples of critical speed, and generation of a multiplicity of unexpected (sideband and harmonic multiple) frequencies which may excite undesirable resonances in system static parts."

The key indicators of this condition are impact tests that provide results for natural frequency much lower than expected and damping higher than expected. Also, vibration data acquired from bearings will contain response at many harmonic multiples of fan speed. All of these conditions applied to the cooling tower fan vibration and modal data.

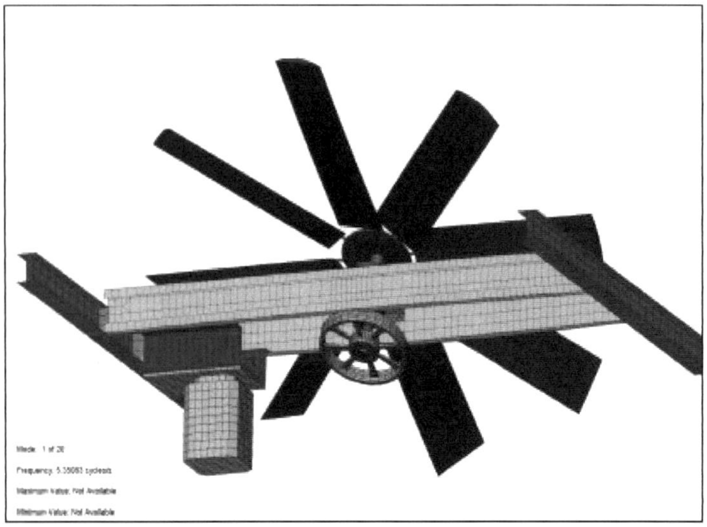

Figure 11: FEA Results for Mode Shape

A finite element model was developed to study the structural dynamic characteristics of the fan rotor relative to the structural supporting system. Figure 11 contains the results of the FEA with the bearing stiffness derated to account for the clearances. This model was developed with full graphical interface with the finite element program that provided immediate verification of the appropriateness of the numerical description of structural-mechanical system being studied. Unlike the FEA in Case Study #1 that required a two-step global-local analysis sequence, this analysis was performed in a single step and was performed using a desktop personal computer versus a main-frame computer.

The key factor in this analysis was approximating the equivalent stiffness of the bearing, accounting for the clearance therein. There are three components of motion/deformation that must be accounted for as the shaft vibrates within the bearing; Δ_c is the displacement within the clearance, Δ_h is the displacement due to Hertzian deformation of the roller elements and races, and Δ_e is the displacement due to the elastic ovalization of the bearing. If these are known, then an equivalent elastic spring can be used. The approximated magnitude of this equivalent spring would be $k = F/(\Delta_c + \Delta_h + \Delta_e)$. The tolerance range for the bearing type was a known quantity, and provided upper and lower bounds for the displacement within the clearance. The natural frequency of the rotor was known from the impact testing, and since the modal mass of the fan was a known quantity, so was the approximate equivalent stiffness. This left the elastic deformation of the bearing ($\Delta_h + \Delta_e$) and the force as unknowns.

Similar to a dynamic balancing procedure, a trial weight was added to the fan wheel, and the vibration response (magnitude and phase) were measured. This represented the response of the system to a combination of the original unbalance in the fan plus the added unbalance provided by the trial weight. A vector subtraction of this measurement from the response prior to the addition of the trial weight provides an estimate of the response to the known dynamic force added to the system by the trial weight.

The mode shape in Figure 11 above was calculated for the upper bound of the natural frequency (5.35 Hz) which would occur for a bearing with a lower amount of clearance. The estimated natural frequency for a bearing with a larger clearance would result in a lower natural frequency. Thus, the bearings that were used for these fans were most probably near the upper bound of clearance for its type. The mode shape of the rotor in a direction perpendicular to the belt tension was nearly rigid body.

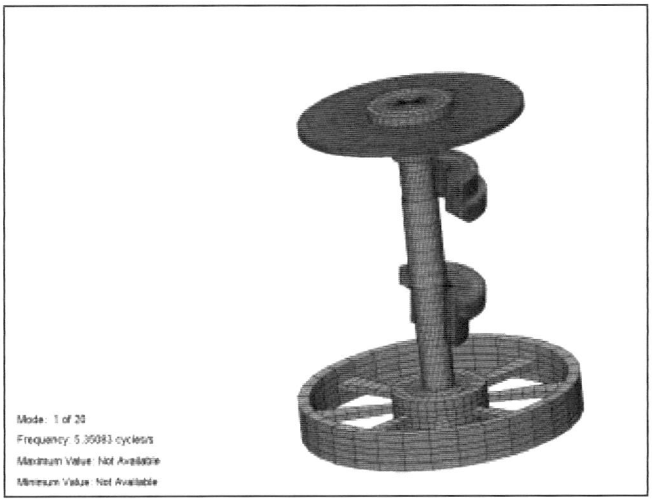

Figure 12: FEA Cutaway View for Rotor Mode

Figure 12 shows a cut-away view of the same mode shape. There is very little flexure in the shaft. The stiffness in this spring-mass model is the non-linear response of the bearing and the support structure. When excited, this natural frequency results in significant back-and-forth sway of the rotor that induces secondary dynamic forces in the belts. Fatigue under the constant influence of the secondary forces was the root-cause of the frequent premature belt failures. The FEA confirmed that the only way that this natural frequency could be increased was to increase the stiffness of the non-linear bearing.

Both the magnitude of the calculated natural frequency and the mode shape are consistent with that which was obtained from actual test data. Data acquired using a shaft stick supported a conclusion that, in the direction perpendicular to the belt pull, the response of the fan rotor was nearly rigid body. The finite element model predicted that the mode shape would be dominated by motion within the bearing and that the participation of the support structure would be minimal.

Once the FEA model was correlated to field data, it could be used as a digital twin for the actual fan. The effectiveness of potential correction action plans could be evaluated using the numerical twin, prior to implementation. The corrective action plan that was chosen consisted of replacing the bearings with a design using a taper lock fit and reducing the clearance in the bearings from a C3 fit to a precision C2 fit.

CASE STUDY #3:

Figure 13: Photo of Fan Failure

In this case study, finite element analysis was necessary to predict the stress stiffened response of a fan rotor. An overhung single-wide, single-inlet (SWSI) fan experienced two catastrophic failures. Figure 13 is a picture of the second failure, in which all the blades and the side plate separated from the backplate of the wheel. The plant at which this fan operated had another unit of identical design directly adjacent thereto. This other fan operated without incident. Although the rotor (shaft and wheel) design of the two fans were identical, there were several interesting differences in the installation. The fan that experienced the failures was driven by a turbine operating at 1,600 rpm, while the fan that did not have any problems was driven by an electric motor operating at 1,600 rpm. The dynamic forces produced by a fan at 1,800 rpm are generally greater than those produced at 1,600 rpm, yet the lower speed fan failed twice and the higher speed did not. This fact pointed to possible speed sensitivity that supported resonance as a probable root-cause.

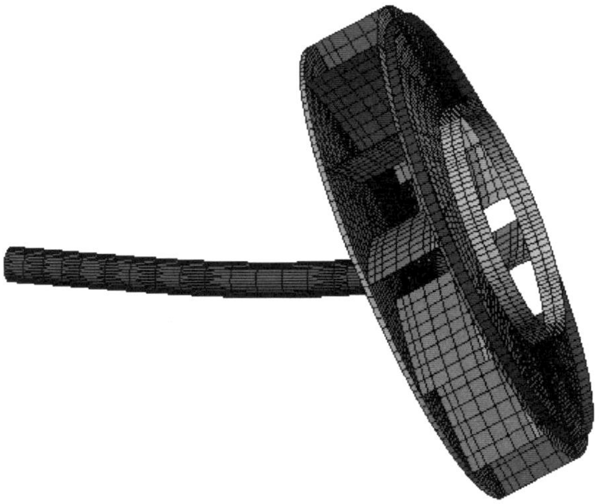

Figure 14: Mode Shape for At-rest Natural Frequency @ 18.2 Hz

A finite element analysis of the fan rotor was done to assist in the probable root-cause analysis. Figure 14 shows the mode shape for a natural frequency calculated as 18.2 Hz. This mode is best described as the principal flexural mode of the rotor. For SWSI fans, this mode always consists of a combination of the flexure of the shaft with the flexure of the backplate of the wheel [7]. Fan rotors of this type will be subjected to large centrifugal stresses, and to a lesser extent gyroscopic effects that will increase the effective stiffness of the wheel when it rotates. The natural frequency calculated by the FEA in Figure 14 predicted the “at rest” natural frequency of the rotor and did not include these stiffening effects.

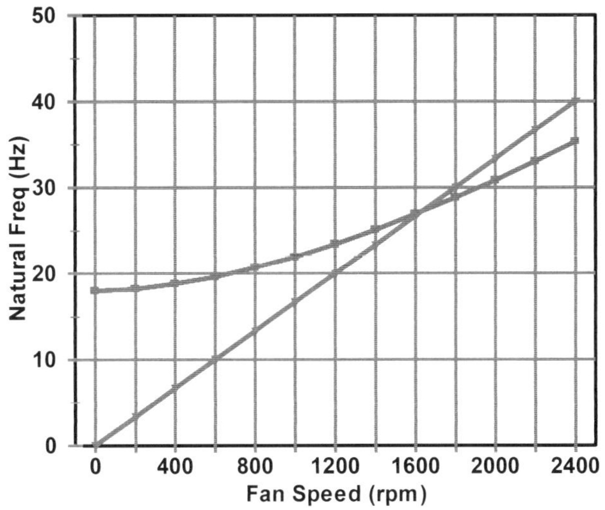

Figure 15: Natural Frequency versus Speed

The finite element method is capable of determining the stiffening effect of centrifugal stresses on the backplate of the fan wheel. As the backplate becomes stiffer, it limits the wobble motion of the wheel, thus increasing the natural frequency. Figure 15 is a curve that shows the change in natural frequency (blue line) with an increase in rotational speed (red line). Since the major driving dynamic force was related to rotational speed (i.e. unbalance), the FEA predicted that the force would be coincident with the natural frequency at an operating speed of 1,600 rpm. At a higher speed of 1,800 rpm, there is a small separation margin between natural frequency and dynamic force. The FEA provided insight into the reason that the lower speed fan failed and the higher speed fan did not.

Since the FEA model provided a good correlation to field data, it could be used as a digital twin for the actual fan. The effectiveness of potential correction action plans could be evaluated using the numerical twin, prior to implementation. The corrective action plan that was implemented included stiffening of the backplate of the fan wheel. The shaft remained unchanged.

The finite element model predicted that the "at rest" natural frequency would increase from 18.2 Hz to around 22.5 Hz. An impact test performed on the modified fan rotor (Figure 16) confirmed the accuracy of the FEA prediction using the digital twin. The FEA model was revised to reflect the changes for this modified replacement design to estimate the effects of centrifugal stiffening during operation.

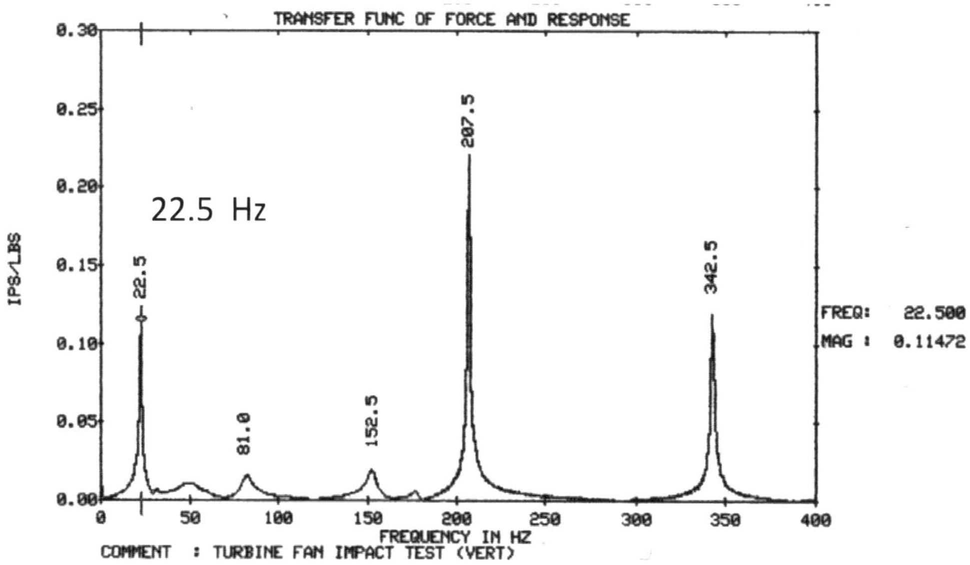

Figure 16: Natural Frequency Test Results for Fan Rotor

Figure 17 is a curve that reports the change in natural frequency (blue line) with an increase in rotational speed (red line) for the replacement design. At the 1,600 rpm operating speed of the turbine-driven fan, the blue line (natural frequency) is separated from the red line (dynamic force) by a sufficient margin.

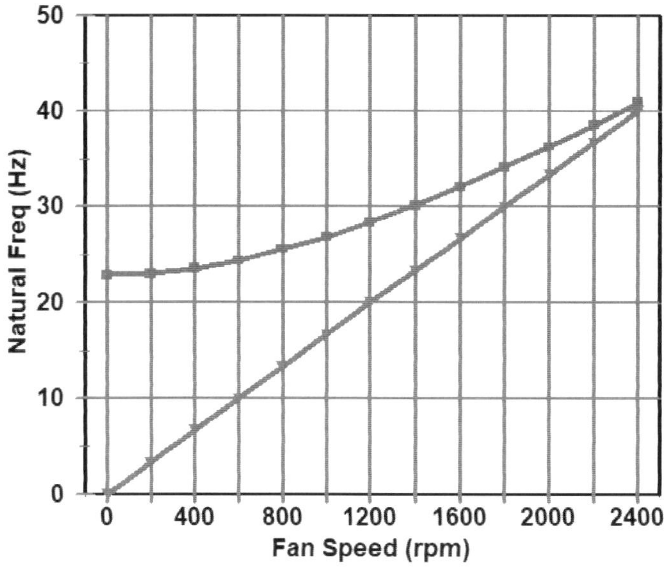

Figure 17: Natural Frequency versus Speed of Modified Rotor

References:
[1] Wohler's Experiments on Strength of Materials, *Engineering*, August, 1967.

[2] J. A. Graham (Editor), *Fatigue Design Handbook*, Advances in Engineering Series (Volume 4), Society of Automotive Engineers, Warrendale, PA, 1968.

[3] R. L.Wetzel (Editor), *Fatigue under Complex Loading*, Advances in Engineering Series (Volume 6), Society of Automotive Engineers, Warrendale, PA, 1977.

[4] R. J. Sayer, *"Structural Dynamics of Centrifugal Fans"*, Proceedings of the National Technical Training Symposium and 34th Annual Meeting of the Vibration Institute, Chicago, IL, June, 2010.

[5] R. J. Sayer, *Mechanical Vibration & Structural Dynamic Analysis of Centrifugal Fans*, Proceedings of the 1st Middle East Conference of the Vibration Institute, Sharm El-Sheik, Egypt, March, 2018.

[6] F. F. Ehrich, "*Handbook of Rotor Dynamics*", Krieger Publishing Company, Malabar, Florida, 1999.

[7] R. J. Sayer, *Rotor Dynamic Considerations of Single-Inlet Centrifugal Fans*, Proceedings of the 3rd Asia Turbomachinery and Pump Symposium, Virtual Conference, February, 2021.

Biography:

Robert Sayer, PE is the Past President of the Vibration Institute and owner of Applied Structural Dynamics (Westerville, Ohio). He has over forty four years of experience as a mechanical and structural engineering consultant to a wide range of industries including pulp & paper, mining & metals, fossil-fuel & nuclear power generation, petrochemical, food processing, glass manufacturing, and aerospace & defense. Mr. Sayer holds engineering degrees from Ohio University (BSCE), Cleveland State University (MSIE), and Purdue University (MSCE). He is a regular contributor at technical conferences, instructor for the Vibration Institute's CAT IV training, and has published over 50 technical articles and co-authored 2 chapters in Art Crawford's "The Simplified Handbook of Vibration Analysis".

MITAGATION OF MACHINERY DISTRESS AT SITE BY UNCONVENTIONAL METHODS – WHERE TIME WAS OF ESSENCE

Author: Mantosh Bhattacharya

VIBRATION INSTITUTE ANNUAL TRAINING CONFERENCE & EXPO 2021
MFPT ANNUAL CONFERENCE

2021 CONFERENCE PROCEEDINGS

MITAGATION OF MACHINERY DISTRESS AT SITE BY UNCONVENTIONAL METHODS – WHERE TIME WAS OF ESSENCE

Mantosh Bhattacharya

Petrofac International UAE

Mantosh.b@petrofac.com

Abstract

During commissioning and start up to normal operating speed of a gas turbine driven compressor, high 1x vibration was observed and found that casing shifting as rare and abnormal phenomena. As there was no plausible means to correct the piping system due to project handover scheduled, a diligent action was taken after detailed analysis of the problem and findings to create an unbalance the compressor rotor to nullify the 1x vibration. After implementation, the turbo compressor was handed over for operation.

With API 673 complaint twelve numbers of centrifugal fans, balanced, tested at OEM works, all fans showed up higher vibration than ISO 10186-3 acceptance criteria with a high 3x axial component. After detailed analysis of all probable causes, in-situ single plane balancing was carried, and vibrations values got lowered.

Keywords – Thermal effect; trim balancing; wobble mode; umbrella mode; single plane balancing

Introduction –

Sometimes due to lacunae in installation or during manufacturing which are overseen can cause rotating machinery anomaly such as high vibrations. Due to tight project schedule, some well deliberated decisions and actions are taken. These two case studies discuss this topic where all risk and rewards were weighted.

Case 1- Unbalancing a balanced rotor.

3 numbers of mission critical gas turbine driven centrifugal compressors handling hydrocarbon gas were procured from a well renowned OEM who had very large fleet record of delivering such packages.

During detailed engineering design and execution all required points mandated by API 617 and other associated standards were followed. This included lateral and torsional rotodynamic analysis the compressor rotors were balanced as mandated in API 617.

Further the flexible coupling units were also balanced as mandated in API 671 and OEM internal standards. In addition, sensitivity analysis for unbalance and verification at OEM test bed was carried out as per prescribed spin test (Mechanical run test) procedure in API 617. The contract coupling was used in spin test of centrifugal compressor. Based on outcomes various analytical studies, references with respect to process parameters of other executed projects, it was mutually agreed that full speed full load complete unit test at OEM works can be waived.

All protocols mandated by OEM manual, API RP 686 were followed for correct installation and alignment. As pre-commissioning activity low pressure low speed (above 1st critical speed of compressor) recycle clean up run was carried out and machine behaved as per expectation in terms of vibration.

However, during start up to operating speed with process hydrocarbon gas, of a gas turbine high 1x vibration was observed as the machine was being ramped up.

It was confirmed by reviewing the documents that compressor rotors were balanced as per required standards and proven methods by OEM (Original equipment manufacturer). The machine was tested as per API standard without any residual technical glitch.

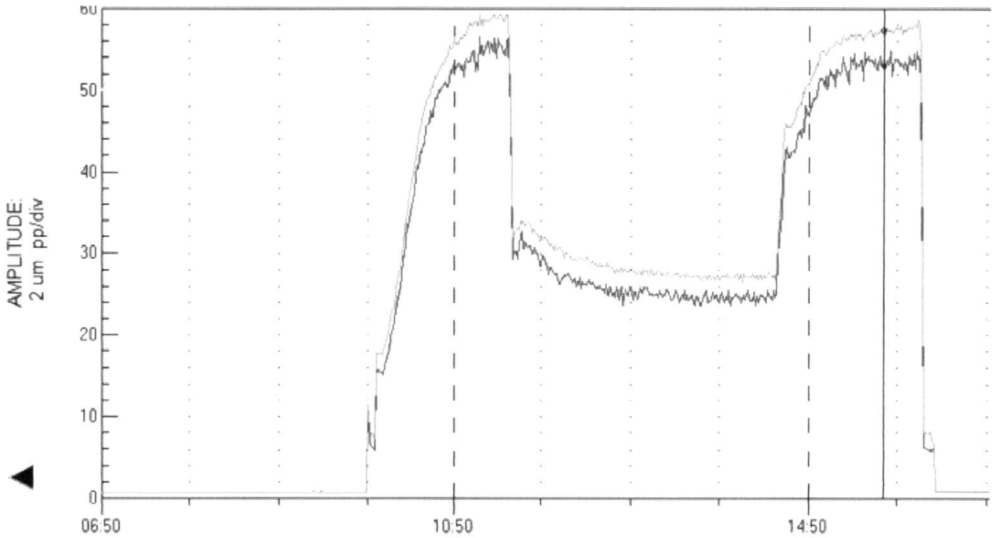

Fig 1 High vibration above alarm value during start and showing speed dependent.

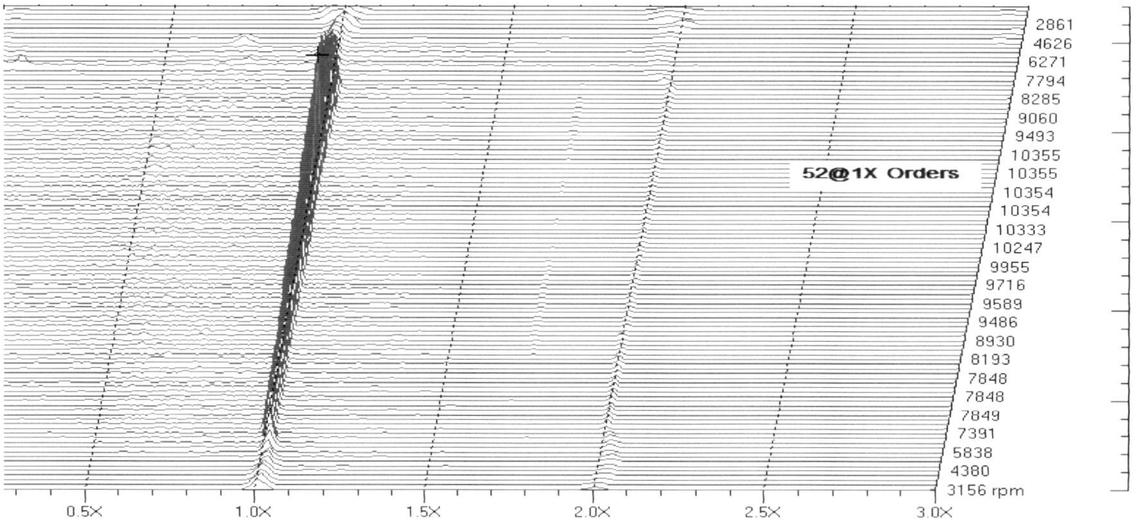

Fig 2 High 1x vibration at compressor DE side

The first work at site was to rotate the coupling hub by 180 degree to nullify the residual unbalance in any on higher side. This trial did not give expected results After checking the bearings for dislocation and other possible causes of 1x vibration at site, attention was focused for a possible misalignment of compressor rotor inside the bearing pad centerline.

Multiple dial gauge indicators were installed on foot of compressor casing position (fig3) and found that casing was shifting from its original position as the machine was being ramped up and gas temperature goes. However, this phenomenon was not observed at low pressure low speed recycle clean up run. It was understood that thermal elongation of a pre-stressed piping was causing compressor casing to pull [fig 4]. The alignment readings and casing strains as measured during installation were checked again and found to be within IOM manual of compressor.

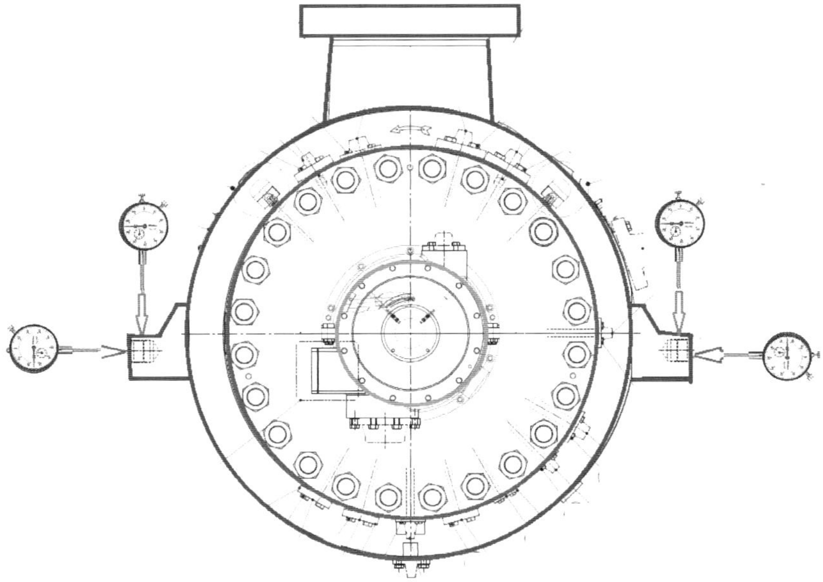

Fig3 – installation of dial gauges to measure casing movement during operation.

It was very sensible to believe that there were some lacunae while installing the piping and somewhere the thermal elongation was muted by using stoppers or struts.

As there was no plausible means to correct the piping system due to paucity of time as complete units were supposed to be handed over to end user, a diligent action was to be taken.

After multiple startups, it was found that only 1x amplitude is dominating at DE side and it happened always after a certain speed and gas temperature. The location (phase) of 1x was carefully found from data plots. It was suggested that to mitigate the 1x amplitude multiple iterations were carried out to create a vectorially opposite force which can lower down the 1x amplitude.

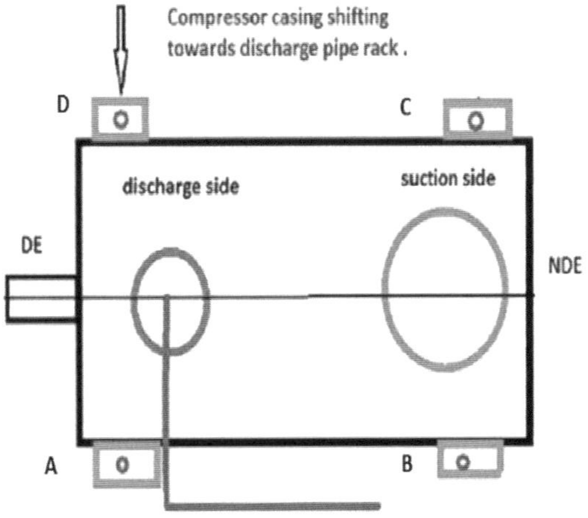

RPM		7500	8500	10000
Casing Temperature		63 deg C	75 deg C	90 deg C
Casing Movement in mm				
A		0.16	0.22	0.45
B		0.015	0.065	0.07
C		0.08	0.06	0.07
D		0.1	0.03	0.04
Shaft Vibration µm - pk-pk	1x			
DE (A-D)			26	33
DE (A-D)			31	50
NDE (B-C)			9.5	12
NDE (B-C)			10	11

Fig4 – corelating Casing temperature, movement, and vibration

On high-speed equipment and/ or lightweight equipment, it may not be possible to provide a balanced coupling to meet the requirements due to the inherent errors introduced when balancing a coupling on a balancing machine. When this occurs, the coupling manufacturer does provide a means where weight can be easily added to the coupling. Then the coupling can be balanced by trial and error on the equipment itself. Couplings can be provided with balancing rings that have radial setscrews or tapped holes so setscrews or bolts with washers can be added or subtracted. There are also other means that can be used.

Based on standard calculations, polar plots and OEM in house proprietary simulation tool, amount of unbalance and position was decided. After implementation of adding the weight at coupling, the turbo compressor was handed over for operation with no high vibration and high 1x component. [sample fig 5 for understanding] The vibration remained well below alarm limits throughout complete operational regime.

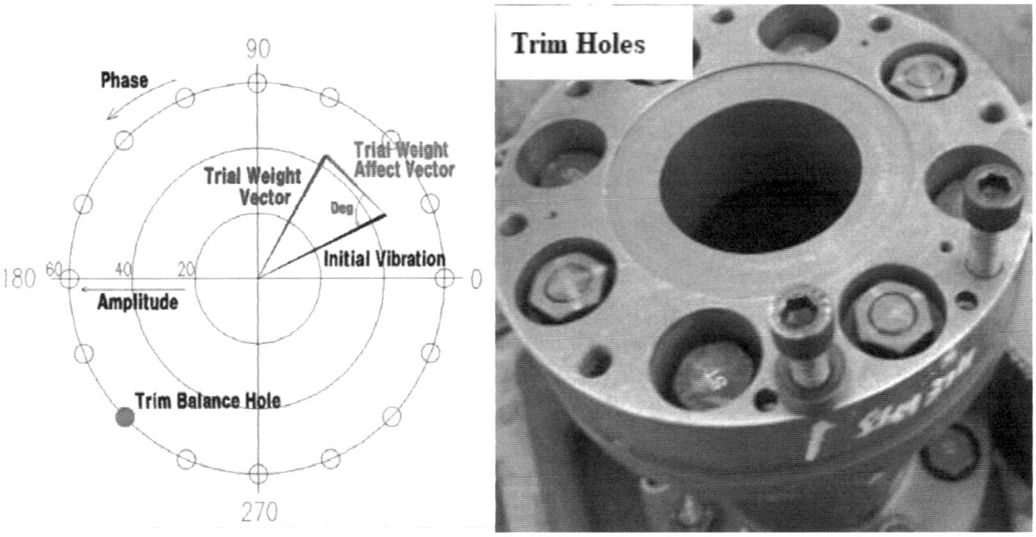

Fig 5–In-situ trim counter balancing at coupling hub. [Courtesy Inertance Korea]

Case 2 – Balancing the balanced rotor.

Twelve API 673 compliant mission critical fans were supplied by a well renowned OEM who had very large fleet record of delivering such packages as air assisted flares and incinerator services in refinery. The fans were between bearing design. During detailed engineering design and execution all required points mandated by API 673 and other associated standards were followed. These included lateral and torsional rotor-dynamic analysis the fan rotors were 2 plane slow speed balanced as mandated in API 673 and as per OEM standards. The fans were tested as per relevant API and AMCA standards at OEM works. The vibration spectrum were healthy and overall vibration levels were compliant to API.

However, when started at site, all blowers were showing high vibration (much higher than acceptance criteria, sometimes reaching alarm level) with a dominant peak of 3x axial along with other regular harmonics. The fans are suspended by double-row spherical roller bearings. The bearings are mounted on a straight shaft.

Multiple investigations were carried out to identify and mitigate the issue to lower down overall vibration within limits as mandated in ISO 10816-3 as operating range was beyond classification of ISO fan vibration limits. The location and possibility of misalignment causing 3x vibration were checked and bearing housings were aligned in terms of shaft centerline. Additionally, the centerline of ……. Site personnel checked the thrust + radial bearing (double row spherical roller bearings) which has good misalignment tackling capabilities, support system for any looseness, coupling hub and fan back shroud face TIR. The TIR distribution throughout the periphery found to be in range of 15-20 mm. It is to be noted that Low frequency axial shuttling of 2 pole motor shaft was observed however this was not related to high axial vibration. The coupling manufacturer confirmed that couplings are designed to counter axial force and shuttling due to mag. Center of the 2 poles motor.

			CPM	2980	5960	8940	10430	11920	16125	14900	17880	25095	35850
Unit	IGV 65% and BLOW OFF VALVE FULL OPEN	OVERALL READING IN RMS	UNIT	1X	2X	3X	3.5X	4X	4.50	5X	6X	7X	10x
Blower A	BLOWER 1 DE H	5.687	MM/SEC	3.725	NS	3.744	NS	0.655	0.693	0.719	NA	NA	NA
	BLOWER 1 DE V	5.77	MM/SEC	2.66	NS	4.237	1.24	0.766	NS	NS	NA	NA	NA
	BLOWER 1 DE A	7.357	MM/SEC	1.782	NS	4.84	3.717	1.465	NS	NS	NA	NA	NA
	BLOWER 1 NDE H	3.612	MM/SEC	2.411	0.526	NS	NS	NS	0.571	NS	1.123	NA	NA
	BLOWER 1 NDE V	2.567	MM/SEC	2.013	NS	NS	NS	NS	0.516	0.32	0.338	NA	NA
	BLOWER 1 NDE A	2.333	MM/SEC	0.982	0.618	1.648	0.354	0.279	NS	0.355	NS	NA	NA

Fig 6– Fan vibration readings

4x mode vibration peak is given by the 4 thrust fins place on the backside of the impeller. The scope of the fins is to reduce the axial thrust on the fixed bearing. This vibration is unavoidable and not dangerous for the fan integrity.

A "classic" condition that will produce axial vibration on a center-hung fan is the shaft bent/bowed between centers. The phase at the bearings taken axially (at the top of the housing) will be 180 out which was not the case after repeat checking at site installation. In one of units the fan was removed insert half key into shaft and unit was started to quickly collect data. The peaks of high axial vibration were not found.

There was no explicit relation found between bearing fault frequencies and 3x vibration.

To investigate further, FE analysis of impeller wheel and fan support structure were carried out.

It was found that 3x axial was coinciding with zero circle, zero diameter mode of impeller. It was also found by FE simulation of the blower structural frame that 3x is having a coincidence with one of resonant mode of base frame as well. With extended FE simulation of impeller dynamic modes, it was concluded that it is difficult to shift the mode with present construction of impeller. Addition of ribs at back-shroud of impeller (extended expeller blades) did not seem to shift this mode. Simulation of adding mass inside the fan support structure and adding ribs did not seem to shift the 3x response.

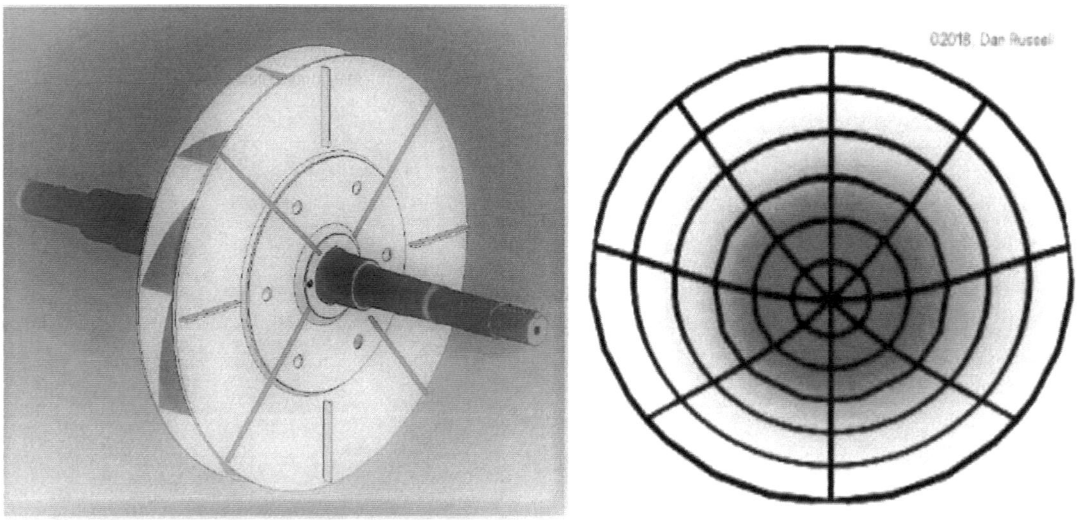

Fig 7– addition of ribs at back shroud and umbrella mode.

A detailed study and understanding the vibration pattern was carried out to identify whether 3x was purely axial or coupled. Normally in standard belief, balancing a wheel does not render a positive result on axial vibration. Vibration pattern was then looked like a wheel wobbling pattern and was decided to refine the balancing grade. Usually, the wobble is the result of distortion when the wheel is fabricated (welded). It is to be noted that during manufacturing of impeller the centerline of orthogonal plane of impeller does not exactly match with centerline of rotational axis of rotor. A welded construction, like any fan of any manufacturer where a certain warping is not avoidable and may cause a wobbling type of effect further accompanied with umbrella mode. Due to standard method of fabrication of large impeller (1000 mm diameter with 220 mm width) with duplex steel, a certain amount of runout always remains. This is known as wrapping. Due this effect both planes of impeller shroud cannot be exactly orthogonal to shaft centerline. This causes a wobbling type of effect. This wobbling type of effect was causing excitation of zero diameter mode and causing 3x axial vibration. The blue line shows desired motion plane of impeller and red line shows the wobbling mode in figure below –

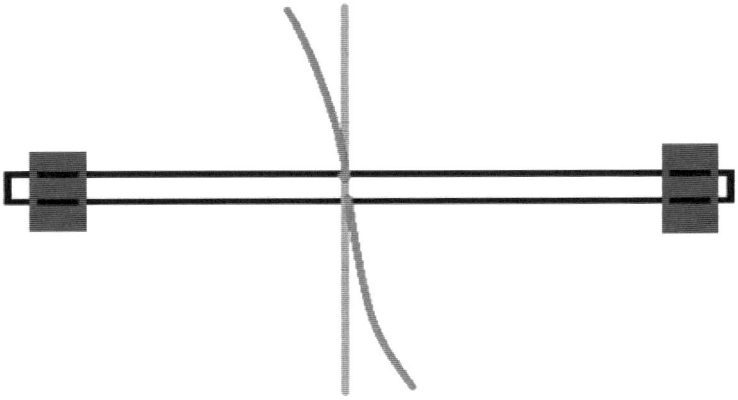

Fig 8- Impeller wobble mode

For overhung fan wheel with wobble, typically single plane balance method is used with weight on intake shroud. This tends to have a better effect on axial vibration than putting weight on backing plate. As the fan was center-hung and front wheel was not accessible location of correction point was different.

The issue was considered as is labeled a disc-wobble resonance based on findings. This resonance involves flexing of the impeller support disc and the pitching inertia of the impeller wheel. As there was no scope to modify the center plate based on construction, it was decided that vibration was to be reduced by single plane field balancing with a controlled manner with accurate amplitude and phase data. It is to be noted that Two balance planes are required when there are two factors involved the disc wobble resonance and shaft bending mode.

The in situ single plane balancing was carried out on blower impellers which lowered down the vibration of blowers within ISO acceptance criteria.

Conclusion

The solutions provided in these case studies do not replace the correct engineering practice of installation and construction / fabrication of machinery. This includes proper support, expansion loops, flexibility of piping for thermal expansion and lower axial runout of fan impeller, proper thrust balance design and proper damping by fan support structure.

References –

1. Couplings - Balancing Tutorial & "New Developments in Gas Turbine Couplings" by Joe Corcoran & Christian Wolford – ATPS 2016
2. The Application of Flexible Couplings for Turbomachinery- Mancuso, Jon R.; Gibbons, C. B. (Barney); Munyon, Robert E. -18th Turbomachinery symposium Texas A&M
3. Field Balance (Trim Balance) – Notes from INERTANCE Inc. Seoul, Republic of Korea

4. Rotor Balancing Tutorial -Ray Kelm, P.E., Dustin Pavelek, P.E.& Walter Kelm, 45th Turbomachinery symposium Texas A&M
5. AHRI Guidelines mechanical balance of impellers for Fans
6. Field Balancing in the Real World: Part 2 CSI Reliability Week October 4-8,1999 Bob Bracher, Plant and Equipment Division and Clark Surrett, Plant and Equipment Division
7. Unique fan vibration problems; their causes and solutions - Donald R. Smith, Harold R. Simmons Southwest Research Institute San Antonio, Texas
8. Decrease axial vibration generator with balancing Guillaume Christin, Nicolas Péton2 GE Measurement & Control,
9. Structural Dynamics of Centrifugal Fans by Robert J Sayer President VI
10. Vibration analysis of centrifugal fans by Robert J. Sayer,
11. Rotor design consideration to prevent impeller and premature bearings failures in Centrifugal fans - Robert J. Sayer, MFPT conference proceedings 2019.
12. Centrifugal fans: using vibration analysis to detect problems - James David Kesler - Technical Associates of Charlotte, P.C.
13. Acoustics and Vibration Animations - Daniel A. Russell, Graduate Program in Acoustics, The Pennsylvania State University

Acknowledgments –

Author tenders his sincere gratitude to Mr. Ganesh Kumar Head of Mechanical department and Mr. Viljoen Christo, Head of integrity assurance Petrofac International for reviews and suggestions. Author is deeply thankful to Mr. Robert Sayer of PE, past president of the Vibration Institute for clarifying some issues related to fan vibration.

The author tenders sincere thanks for support from VI review committee for review and accepting this paper for presenting in VIATC 2021 annual conference.

USING MOTOR CURRENT SIGNATUTE ANALYSIS TO DETECT THE NUMBER OF BROKEN ROTOR BARS IN INDUCTION MOTORS

Author: Mohamed Diaaeldien Hussien Wahba

USING MOTOR CURRENT SIGNATUTE ANALYSIS TO DETECT THE NUMBER OF BROKEN ROTOR BARS IN INDUCTION MOTORS

Mohamed Diaaeldien Hussien Wahba
Ezz Steel Industries
Economic Industrial Zone Northwest Gulf of Suez km 44 Suez/Hurghada Road Ain Sokhna – Suez, Egypt.
Wahba@it-enterprise.com

Abstract: Cracked/Broken rotor bar(s)is one of the commonly encountered induction motor faults that may cause serious damage to the motor if not detected timely. Past efforts on broken rotor bar diagnosis have been focused on both vibration analysis and motor current signature analysis (MCSA). These methods require accurate slip estimation along with high resolution spectral data which enables the analyst to precisely judge the severity of the problem, and confidently pull out the motor to the workshop with no doubt.

This paper will not only focus on collecting current signature, but also will deeply cover the analysis part to determine the number of broken bars if exist, and explicitly evaluate the severity level and needed corresponding corrective actions. This will be clearly shown through two real case studies for two (1.2 MW) induction motors in a large integrated steel plant.

Keywords: Broken Bars, Current Signature, Cracked Bars, Induction Motor, MCSA, Slip Frequency, Steel Industry, Vibration Analysis.

This paper presents results from an in-depth investigation of using current signature technique and its impact on the fault detection of rotor bars in induction motors. The analysis is performed based on both vibration data and current signature, where the first indication of anomaly came from vibration Time-domain data. The paper includes two case histories

1.INTRODUCTION:

Induction motors play a significant role in the safe and efficient operation of industrial plants. Usually, they are designed for 30 years fault-free lifetime, but most of them are not available at all times.

Most failures in induction motors can be classified in two main groups: isolation failures and mechanical failures (Botha, 1997). Isolation failures are commonly characterized by stator coils short-circuits, while mechanical faults are commonly associated to rotor or rotor related damage. The most important mechanical failures are: rotor broken bars and rings, bearings damage, irregular gaps (static and dynamics eccentricities), unbalances, etc.

Motor Current Signature Analysis (MCSA) is a condition monitoring technique used to diagnose problems in induction motors. Concept originates from early 1970s and was first proposed for use in nuclear power plants for inaccessible motors and motors placed in hazardous areas, [1]. It is rapidly gaining acceptance in industry today. Tests are performed online without interrupting production with motor running under the load at normal operating conditions, [1]. MCSA can be used as predictive maintenance tool for detecting common motor faults at early stage and as such prevent expensive catastrophic failures, production outages and extend motor lifetime. It can be used as a diagnostic tool and powerful addition to vibration and thermal monitoring (verifying a fault with more than one technology), [1].

A crucial point about motor current signature analysis (MCSA) is that it is sensing an electrical signal that contains current component that are a direct by-product of unique rotating flux components caused by faults such as broken rotor bars, airgap eccentricity, and shorted turns in low voltage stator windings, etc. MCSA can detect these problems at an early stage and thus avoid secondary damage and complete failure of the motor, as reported by Hargis, et al. (1982), Thomson (1984), Thomson and Rankin (1987), and Kliman and Stein (1990). It is true that broken rotor bars will result in a change to the vibration spectrum, but vibration is traditionally sensed at the bearings. And for each motor there is a different mechanical stiffness between the electromagnetic forces caused by broken bars and the position where the vibration is sensed. This adds an additional complexity when attempts are made to quantify the severity of the problem via vibration analysis. Electromagnetic forces are proportional to the flux density squared waveform (Alger, 1965; and Yang, 1981) in an induction motor. Hence, the vibration from unique electromagnetic forces from broken bars, etc., is a second order effect compared to current components directly induced from the specific rotating flux waves. In many cases the fault severity (e.g., number of broken rotor bars) has to be serious before it can be detected by vibration analysis, and even then, the prediction of fault severity is another order of magnitude more difficult, [2].

2.HOW MCSA WORKS:

The technique depends on spectrum analysis of harmonic components in the line current produced of unique rotating flux components caused by faults such as broken rotor bars, air-gap eccentricity and shorted turns in stator windings, etc. The method needs only one current transducer, and it can be in any one of the three phases.

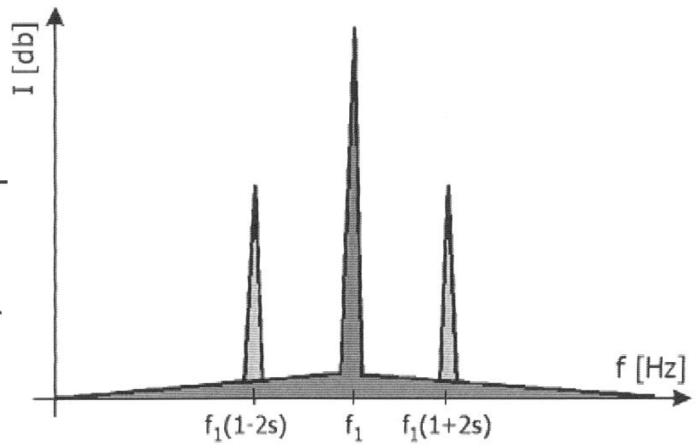

An ideal current spectrum is shown in Fig. 1. The two slip frequency sidebands due to broken rotor bars near the main harmonic can be clearly observed.

Figure 1- Current Spectrum

Usually a decibel (dB) versus frequency dynamic range and to detect the unique current signature patterns that are characteristic of different faults [2, 3].

In the three-phase induction motor under perfectly balanced conditions (healthy motor) only a forward rotating magnetic field is produced, which rotates at synchronous speed,[6] $n1 = f1/p$, where:

f_1: Supply Frequency
p : Pole Pairs

The rotor of the induction motor always rotates at a speed (n) less than the synchronous speed.

$$f_2 = n_2\, p = s\, n_1 p \qquad (1)$$

The speed of the rotating magnetic field produced by the current carrying rotor conductors with respect to the stationary stator winding is given by:

$$n + n_1 = n_1 + n_1 - n = n_1 \qquad (2)$$

With respect to a stationary observer on the fixed stator winding, then the speed of the rotating magnetic field from the rotor equals the speed of the stator rotating magnetic field, namely, the synchronous speed.
Both mentioned fields are locked together to give a steady torque production by the induction motor. With broken rotor bars in the motor there is an additional, backward rotating magnetic field produced, which is rotating at the slip speed with respect to the rotor. The backward rotating magnetic field speed produced by the rotor due to broken bars and with respect to the rotor is:

$$n_b = n - n_2 = n_1(1 - s) - sn_1 = n_1 - 2sn_1 = n_1(1 - 2s) \qquad (3)$$

The stationary stator winding now sees a rotating field at:

$$n_b = n_1(1 - 2s) \qquad (4)$$

Or expressed in terms of frequency:

$$f_b = f_1(1 - 2s) \qquad (5)$$

This means that a rotating magnetic field at that frequency cuts the stator windings and induces a current at that frequency (f_b). This in fact means that (f_b) is a twice slip frequency component spaced $2sf_1$ down from f_1.Thus speed and torque oscillations occur at $2sf_1$ and this induces an upper sideband at $2sf_1$ above f_1.

Thus, twice slip frequency sidebands occur at $\pm 2sf_1$ around the line frequency:

$$f_b = (1 \pm 2s)f_1 \qquad (6)$$

While the lower sideband is specifically due to broken bar, the upper sideband is due to consequent speed oscillation. In fact, several papers shows that broken bars actually give rise to a sequence of such sidebands given by:

$$f_b = (1 \pm 2ks)f_1 \quad , K = 1,2,3\ K \qquad (7)$$

3.CASE HISTORY 1

Fume Treatment Plant – ID Fan Motor # 2, EzzSteel, Suez, Egypt

Case Information:

Case Date	July -2017
Location	EzzSteel Plant, Suez City, Egypt
Division	Melt Shop, Fume Treatment Plant
Motor Brand	WEG
Duty	Special
Motor Power	1.2 MW
Supply Current	50 HZ
HP	1630
Volt	11000
Amp	77.2
RPM	1000
No. of Poles	6
No. of Slots	58

3.1 Analysis Procedures and Interpretation

3.1.1 Routine Vibration Data:

Vibration data at this plant is used to be collected on bi-weekly basis, as this area has been classified as "Critical" due to its importance in the process, safety, and work environment. Emerson, CSI 2130 was used on these measurements, and "MHM" machine Health Management was used in analysis of the data.

3.1.2 Vibration Spectral Data:

Spectral data for this motor was very common for most of vibration analyst, residual rotating looseness presented in 1x turning speed, and its harmonics. According to this information, its not evident to diagnose a critical event in the motor.

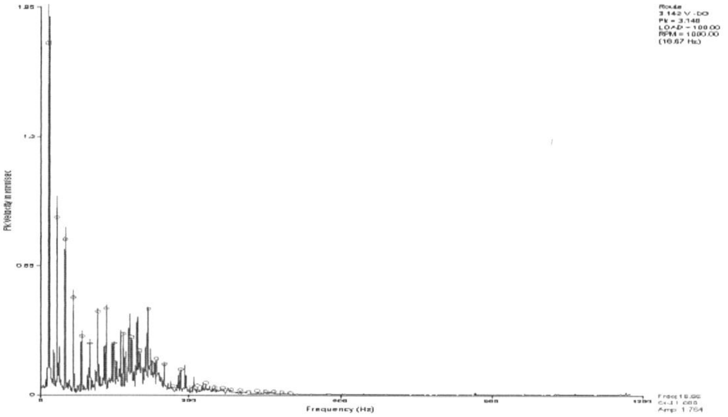

Fig 2- Motor Spectrum shows Some looseness

3.1.3 Vibration Time-Domain Data:

While spectral data doesn't show a serious problem, going through the time -domain data gives more insights for this case. Amplitude modulation is very clear in the waveform, and this is a serious indication that should be considered in our analysis, after all, that's why we see the sidebands.

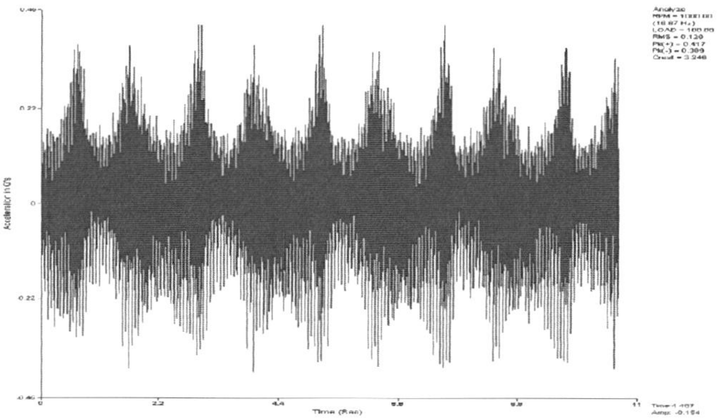

Fig 3- Motor Time-Domain Data

3.1.4 Re-taking Vibration Spectra:

I asked for a high-resolution spectrum with a lower Fmax to be taken as this was not normally collected as part of the route. This spectrum has a Fmax of 100 Hz and with 6400 lines. You can now see that those peaks of 1X, and harmonics we saw are not simple peaks – they are broad because of the sidebands surrounding each of the peaks. With limited resolution,

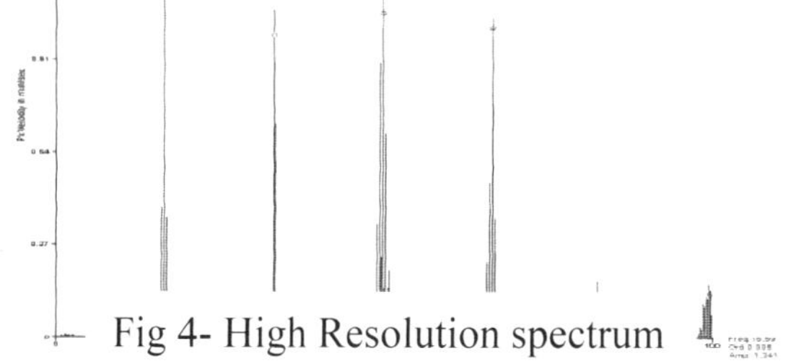

Fig 4- High Resolution spectrum

there is no way to tell that those sidebands are present.

Zooming in at one of these 1xs', then you cab obviously see the sidebands, and consequently detect the frequency of these sidebands. this is defintely the core part of the analysis that led you to the rotor bar problems.

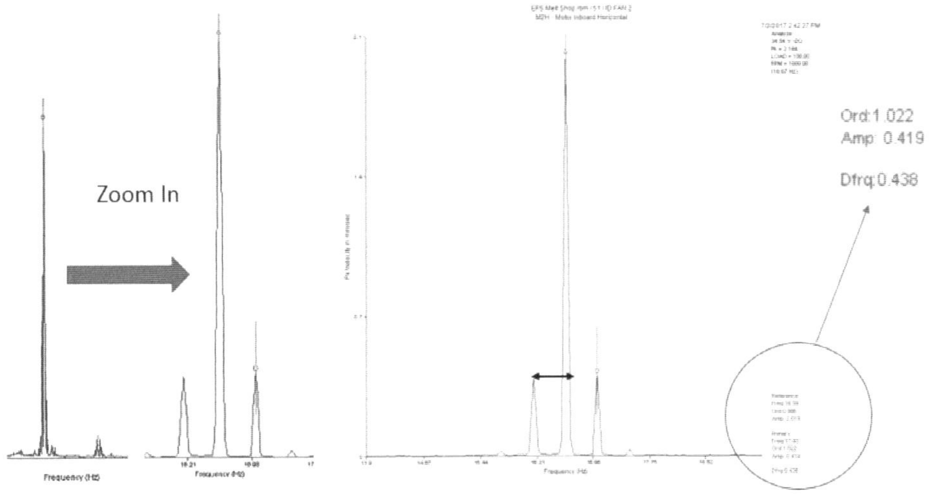

Fig 5- High Resolution spectrum

3.1.5 Spectral Data Interpretation:

Now the clue comes from the sideband frequency, as we see in the spectrum, the difference between the fundamental frequency and the left sideband is 0.438 Hz. This small value will lead us to introduce the terminology of "Pole Pass Frequency". Until further illustration for this term in the following pages, I will introduce the following terms, and calculations:

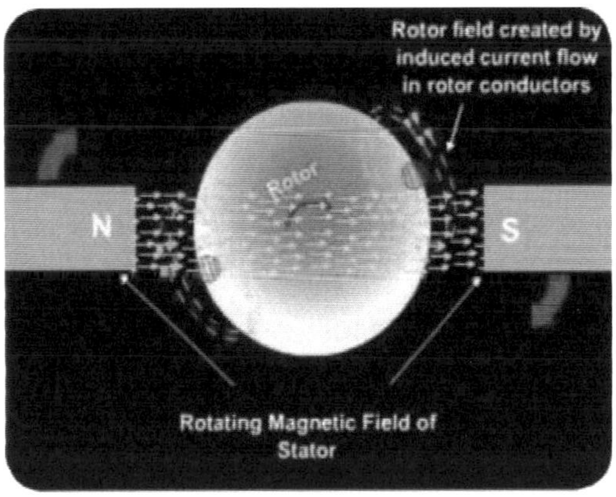

Fig 6- Rotor & Stator Construction

Slip frequency (S):
Difference in speed between the rotating field and the rotor bars.

S = Ns – Nr / 60, Where:
Ns: Synchronous Speed
Nr: Rotor Speed

Calculating Motor No. of Poles (P)
P = 120 X LF / Ns,
LF=50 Hz in Egypt

Calculating Fpp:
Fpp = Slip Frequency (S) X No. of Poles (P)
In our Case

Motor Data:
RPM: 1000 (Name Plate RPM = Ns) , RPM: 995.6 (Rotor Actual RPM by Tachometer)

Fpp = [(Ns-Nr)/60] X [(120XLF/Ns)] = [1000-995.62/60]X[120x50/1000]
Fpp = 0.438 Hz

3.1.5 Time-Domain Data Interpretation:

Again, the difference in the time-Domain between any two peaks gives the same Fpp, which is 0.438Hz, this means that the running speed (1X) is modulated by the pole pass frequency.

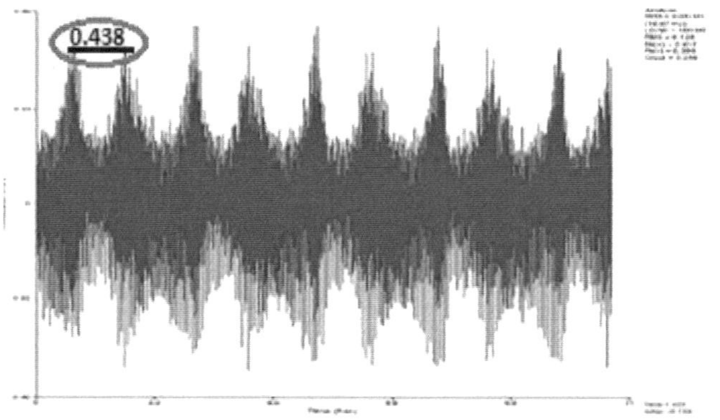

Fig 7 – 1X is Modulated by Fpp

3.2 Fault Mechanism and Motor Current Signature Analysis

3.2.1 Data Collection:

In this case we used: CSI analyzer with CSI Ac Current Probe Model 341B
Input: 600 A AC Max & Output: 0.005 V/A

It can be in any one of the three phases, because the rotating flux waves produced by the different faults cut all three stator phase windings, and corresponding currents are induced in each of the three phases.
The Motor was running and loaded at least 75%. To get reliable data from current signature the motor must be under load. The motor was running at 995 RPM measured by the Tachometer. As long as we need to see the line frequency (50Hz) and the sidebands, we selected the Fmax = 150Hz

Fig 8- Collecting Current Signature Using CSI Ampere Clamp & CSI Analyzer.

3.2.2 Data Interpretation – First Snapshot:

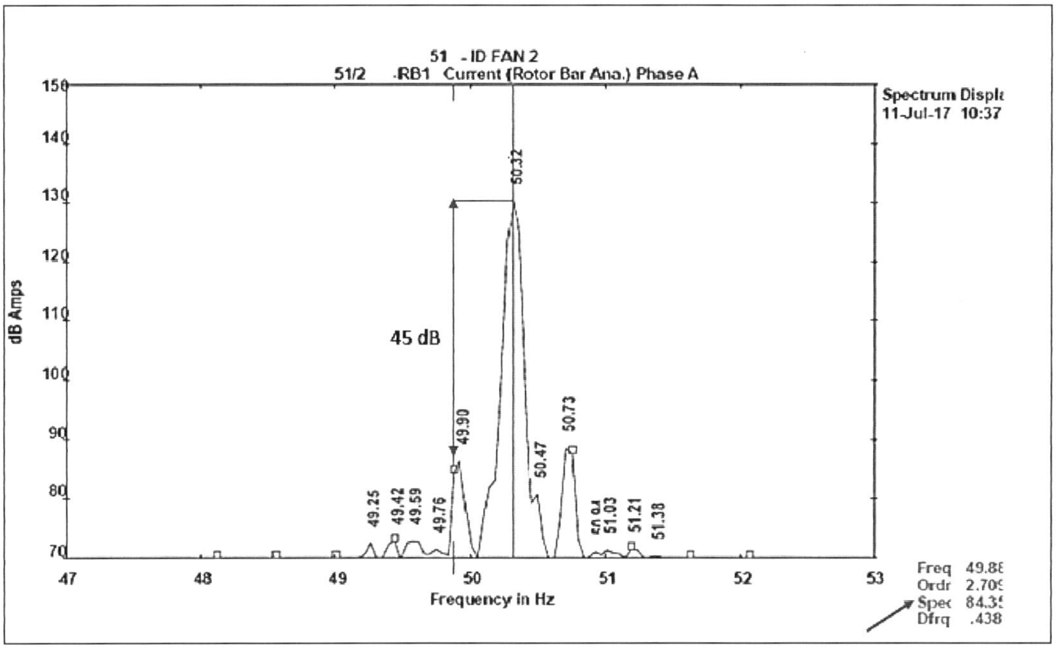

Fig 9- Current Signature frequency domain on 11-July 2017 Shows 45dB difference between line frequency Amplitude and Left sideband Amplitude.

In Fig 9, we can see the line frequency (50Hz) recorded an amplitude 130 dB approx., and two obvious sidebands which reflect the existence and the severity of the rotor bar defect. The amplitude of the left sideband as shown in red data recorded 84.35 dB, this value increase as the defect becomes more severe. In this signature which had been taken on 11-July -2017, the difference between the height of the line frequency 50Hz, and the height of the left first sideband was 45dB, this value is our indicator of rotor bar deterioration.

dB	Rotor Condition Assessment	Recommendations
> 60	Excellent	None
54 - 60	Good	None
48 - 54	Moderate	Trend Condition
42 - 48	High Resistance Connection or cracked rotor Bars	Increase Test Frequency And Trend
36 - 42	Broken Rotor bars will show In Vibration	Confirm With Vibration , Plan Repair , Replace
30 - 36	Multiple Cracked/ Broken Rotor Bars	Repair / Replace ASAP
< 30	Severe Rotor Fault	Repair /Replace Immediately

Fig 10- CAT III, Mobius Institute Manual, dB value, Rotor Condition, and Recommendations.

According to the previous table, the condition of the rotor bars of this motor is at a stage which needs a very close monitoring, the suspicion of cracked rotor bars exists. As a result, the frequency of the measurements has been increased to be twice per week and we introduced thermography activity to detect any expected overheat.

When a broken bar is present within the rotor, current cannot flow through it, and therefore, it can no longer add its share of torque to the rotors load burden. As the broken bar passes under the pole it will effectively reduce the torque of the rotor for the period of time it is under the field pole, in its torque producing position. Since the torque of the rotor is reduced slip is again introduced. As the rotor slows and the slip is increased, the voltage induced into the rotor is also increased, and therefore, the current in the good rotor bars will have to increase to produce the needed torque. This continues until the torque of the motor is restored to equal the torque of the load. When the broken bar is in a position where its contribution is not relied upon (traveling parallel to the lines of flux), the torque of the motor will go up and the opposite effect will be true. This beating is very clear in the time waveform. This analysis method is referred to in several ways - lower side band analysis, pole pass frequency sideband analysis, rotor bar signature analysis, as well as a few others, [4]. Regardless of the name, the current spectrum surrounding the fundamental frequency is closely examined for the relative amplitude difference between the fundamental and what is referred to as the rotor bar sideband frequency (Frotorbar). The amplitude increase at the Frotorbar is due to the added current drawn in the rotor and also from the stator as the good bars attempt to make up for the one defective bar. As the amplitude increases at the rotor sideband frequency, the amplitude difference between the fundamental frequency, and the rotor bar frequency will go down. As the rotor bars break, and the health of the rotor continues to decline, this difference will also decline over time as the sideband frequency becomes more and more dominant in the signal, [4].

3.2.3 Data Interpretation – Second Snapshot:

After few days, we captured a new current signature to figure out the status of the dB value, the measurement conditions were the same and surprisingly the dB values dropped from 45dB to 42dB. This was a serious alarm for the bars condition and the deterioration rate. Moreover, the height of the lower sideband on the left side of the line frequency increased from 84.2 Hz to 90Hz, this indicates the rotor bar defect frequency and how the severity increases by the time. At 42dB, we can expect one rotor bar breakage. At this stage we issued our recommendation to stop the motor and pull it out to the workshop for immediate inspection, but unfortunately due to some production due dates we could not stop this motor. We increased the frequency of measurement either vibration or current signature to be on a daily-basis. We could notice the drop of dB value week by week.

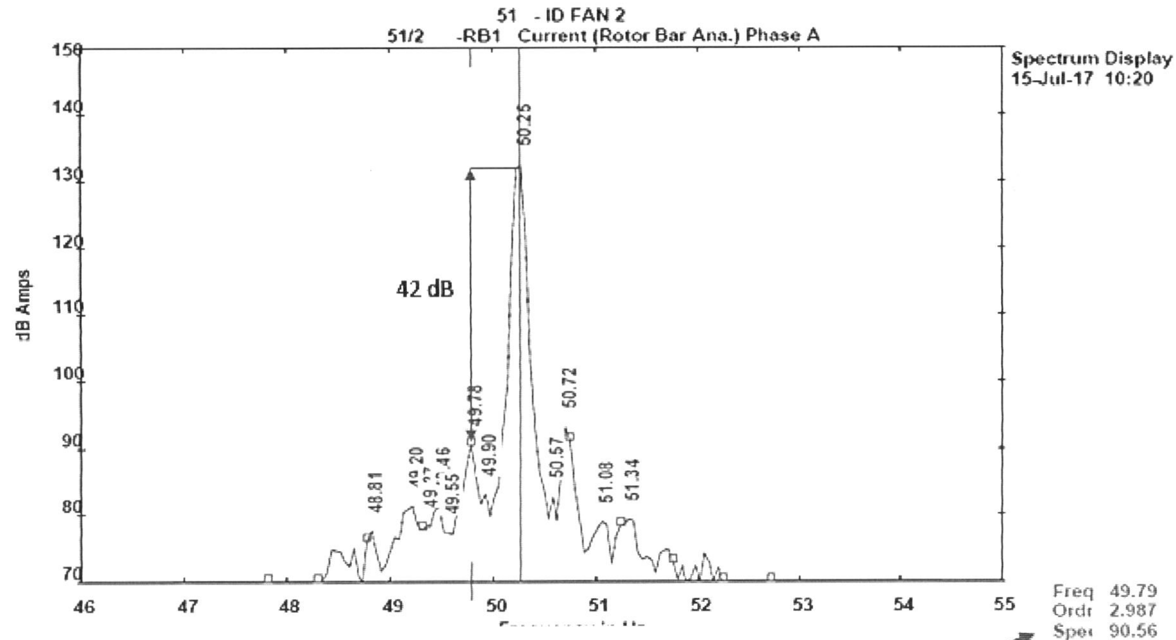

Fig 11- Current Signature frequency domain on 15-July 2017 Shows 42dB difference between line frequency Amplitude and Left sideband Amplitude.

3.2.4 Data Interpretation – Third Snapshot:

On 26th July-2017, the dB value dropped significantly to 34. 3dB.At this stage -even its too late-, we decided to stop the motor and pull it out from the operation and send it to the workshop for inspection. You can clearly hear the humming sound close to motor area, and feel the beating you already see in the waveform.

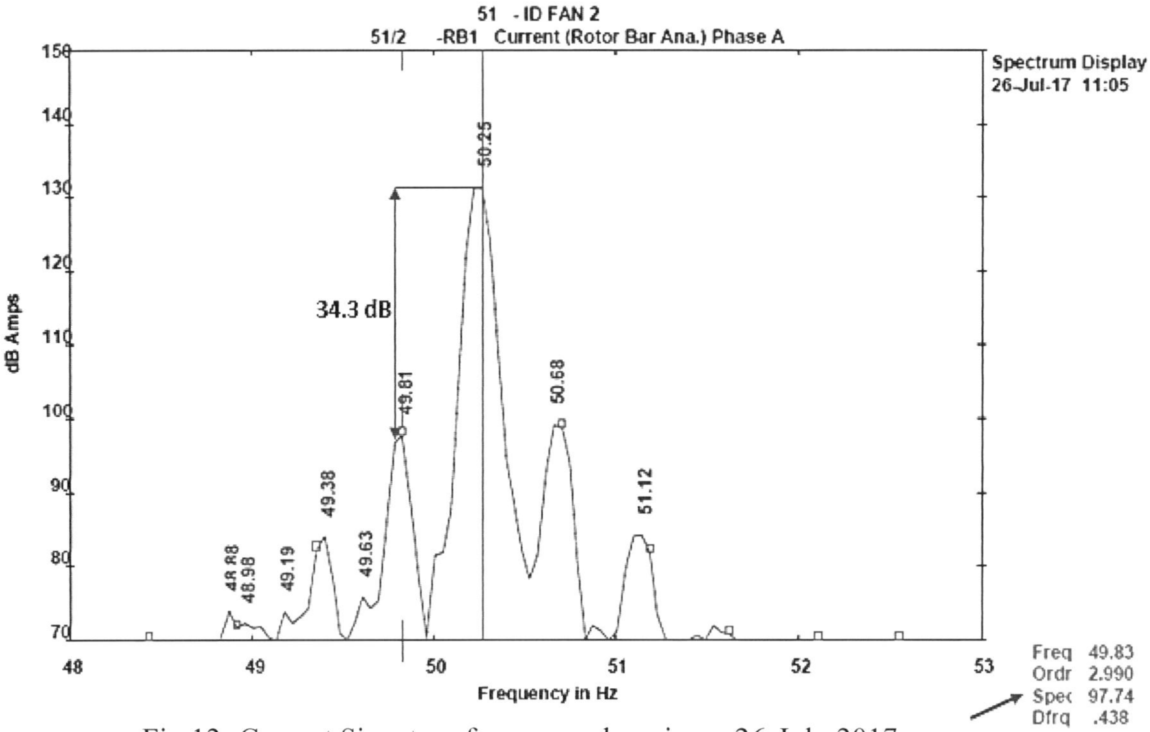

Fig 12- Current Signature frequency domain on 26-July 2017 Shows 34dB difference between line frequency Amplitude and Lower Left sideband Amplitude.

3.2.5 Motor Visual Inspection in The Workshop:

By Pulling the motor out to the workshop and inspect it we found a completely broken bar fallen from its position on the rotor, three cracked bars, and parts of melted lamination.

Fig 13- One completely Broken Rotor Bar in The Motor

3.2.5 Estimating The Number Of Broken Rotor Bar:

$$N = \frac{2R}{(10^{n/20} + 3)}$$

(Thomson and Rankin 1987, [2])

N: Estimate of Number of Broken Bars
R: Number of Motor Slots
n: Average dB difference between upper and lower sidebands and supply component.
P: Pole Pairs

3.2.6 Applying the Equation:

$$N = \frac{2R}{(10^{n/20} + 3)} = \frac{2 \times 58}{(10^{39/20} + 3)}$$

<u>N= 1.24 Bars</u>

It was therefore predicted that there was definitely one bar completely broken. There was also a risk of the broken bar lifting out of the slot with consequential mechanical damage to the stator winding.

Fig 14- The Equation validates the one broken bar existence.

4.CASE HISTORY 2

Fume Treatment Plant – ID Fan Motor # 4, EzzSteel, Suez, Egypt
Case Information:

Case Date	Nov -2019
Location	EzzSteel Plant, Suez City, Egypt
Division	Melt Shop, Fume Treatment Plant
Motor Brand	WEG
Duty	Special
Motor Power	1.2 MW
Supply Current	50 HZ
HP	1630
Volt	11000
Amp	77.2
RPM	1000
No. of Poles	6
No. of Slots	58

4.1.1 Data Interpretation – First Snapshot:

In this case the dB value was 44dB, but we decided at this stage to pull out the motor to workshop to mitigate any further risk, and to avoid any damage as we experienced in the Case 1.

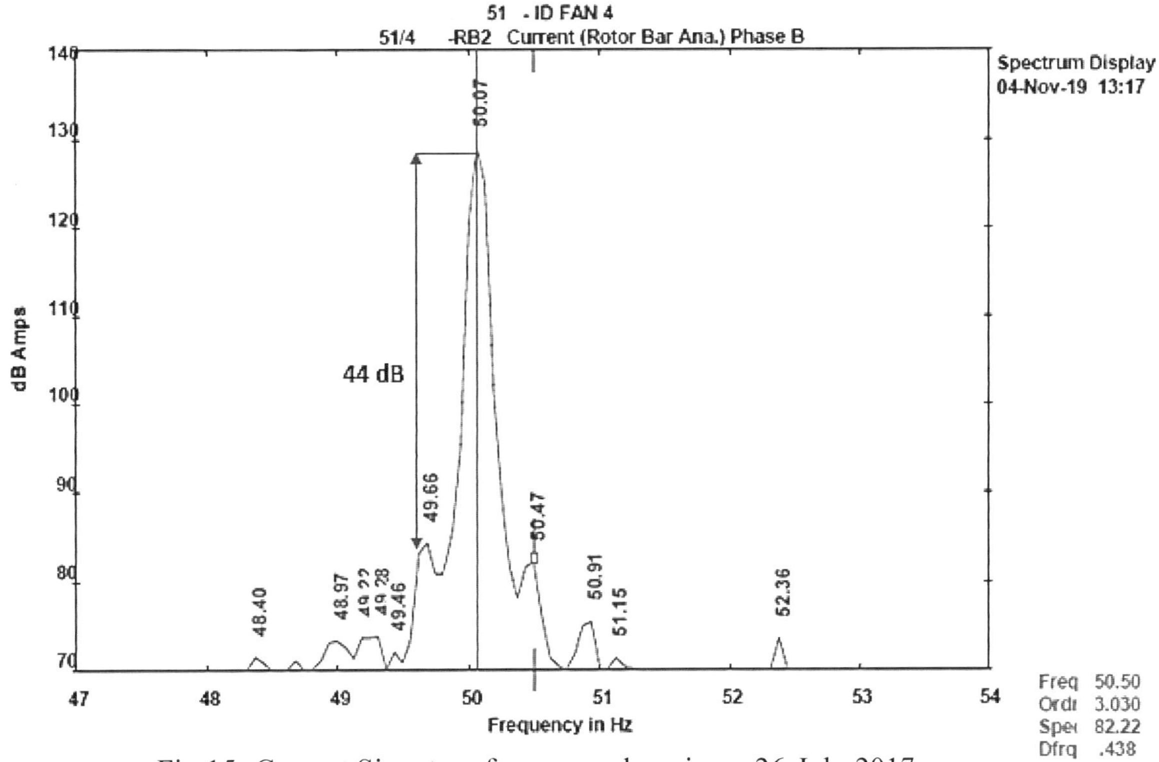

Fig 15- Current Signature frequency domain on 26-July 2017
Shows 44dB difference between line frequency Amplitude and
Lower Left sideband Amplitude.

4.1.2 Estimating the Number of Broken Rotor Bar:

$$N = \frac{2\,R}{(10^{n/20} + 3)} = \frac{2 \times 58}{(10^{44/20} + 3)}$$

<u>N= 0.7 Bars</u>

In this case, we could catch the severity level very early and the equation again proves the prediction of the existence of only cracked rotor bars, an no broken bars.

Fig 16- Two Cracked Rotor Bars

REFRENCES:

[1] A. Korde: On-Line Condition Monitoring of Motors Using Electrical Signature Analysis, Recent Advances in Condition Based Plant Maintenance, 17-18 May 2002, Mumbay.

[2] Thomson, W.T., and Gilmore, R.J., (2003), "Motor Current Signature Analysis to Detect Faults in Induction Motor Drives – Fundamentals, Data Interpretation, and Industrial Case Histories," Proceedings of 32nd Turbomachinery Symposium, Texas, A&M University, USA.

[3] Milimonfared, J. et al., (1999), "A Novel Approach for Broken Rotor Bar Detection in Cage Induction Motors," IEEE Transactions on Industry Applications, vol. 35, no. 5, pp. 1000-1006.

[4] Drew Norman, Reliabilityweb.com, Articles, Diagnosing rotor bar issues with torque and current signature analysis.

[5] Jason Tranter | Founder & CEO, Mobius Institute, cbmconnect.com, electric motor fault,

[6] M. El H. Benbouzid, A Review of Induction Motors Signature Analysis as a Medium for Faults Detection, IEEE Transactions on Industrial Electronics, Vol. 47, No. 5, Oct. 2000.